IoT Enabled-DC Microgrids
Architecture, Algorithms, Applications, and Technologies

Editors

Imed Ben Dhaou

Department of Computer Science
Hekma School of Engineering, Computing and Design
Dar Al-Hekma University, Saudi Arabia
Department of Computing
University of Turku, Finland
Department of Technology
Higher Institute of Computer Sciences and Mathematics
University of Monastir, Tunisia

Giovanni Spagnuolo

Department of Information and
Electrical Engineering and Applied Mathematics
University of Salerno, Fisciano, Italy

Hannu Tenhunen

Faculty of Information Technology and Communication
Tampere University, UC Pori, Finland

CRC Press
Taylor & Francis Group
Boca Raton London New York

CRC Press is an imprint of the
Taylor & Francis Group, an **informa** business

A SCIENCE PUBLISHERS BOOK

First edition published 2025
by CRC Press
2385 NW Executive Center Drive, Suite 320, Boca Raton FL 33431

and by CRC Press
4 Park Square, Milton Park, Abingdon, Oxon, OX14 4RN

Library of Congress Cataloging-in-Publication Data (applied for)

ISBN: 978-1-032-59411-8 (hbk)
ISBN: 978-1-032-59412-5 (pbk)
ISBN: 978-1-003-45457-1 (ebk)

DOI: 10.1201/9781003454571

Typeset in Times New Roman
by Prime Publishing Services

Preface

In 2015, global carbon dioxide emissions reached a staggering 56 gigatonnes (Gt), with over 73% of these emissions in 2020 attributed to the energy sector, as reported by the UN climate change agency. This alarming reality has spurred the urgent need for a drastic action plan to curb greenhouse gas emissions by transitioning from fossil fuels to renewable energy resources, such as wind and solar energies. At the forefront of this transition is the microgrid, playing a pivotal role in seamlessly integrating distributed energy resources into the distribution network.

The roots of the microgrid concept trace back to 1882 when Thomas Edison inaugurated the Pearl Street Station, a decentralized small-scale power system designed to illuminate Manhattan's streets. It wasn't until the mid-20th century microgrid concept when microgrid gained widespread acceptance. Recent advances in semiconductor technology, sensors, and communication technologies have propelled the transition towards intelligent microgrids. The efficiency and resiliency of these intelligent microgrids are achieved by incorporating advanced algorithms, data analytics, automation, and artificial intelligence to optimize energy generation, storage, and consumption. Moreover, DC appliances, the electrification of transportation, the integration of renewable energy sources, and the enhanced efficiency of DC storage units have positioned DC microgrids as a viable alternative to AC microgrids.

This edited volume serves as a compendium of groundbreaking research and insights into various facets of DC microgrid technology. Divided into two parts, the book delves into the Fundamentals and Technologies for DC Microgrid, comprising chapters on Architectures and Technologies, IoT-Based Communication, Blockchain Technology, Photovoltaic Digital Twin, Cybersecurity Perspectives, and AI-Driven Battery State Estimation. The second part focuses on the Design and Optimization of DC Microgrids, exploring topics like hybrid microgrid design for the cement industry, formal methods for microgrids, and a business model perspective on direct current microgrids.

The editors extend their sincere appreciation to the esteemed authors whose invaluable contributions have enriched this edited volume on DC microgrid technology. Gratitude is also expressed to the research center at Dar Al-Hekma

University for providing essential financial support, enabling the realization of this collaborative endeavor. Additionally, the editors would like to acknowledge and commend the diligent efforts of the reviewers, whose insightful feedback and expertise significantly contributed to improving the scope and content of the book.

The editors:

Prof. Imed Ben Dhaou, Dar Al-Hekma University, Saudi Arabia
Prof. Giovanni Spagnuolo, University of Salerno, Italy
Prof. Hannu Tenhunen, University of Tampere, Finland

Contents

Part 1

Fundamentals and Technologies

CHAPTER 1

Architectures and Technologies for DC Microgrid

Imed Ben Dhaou [a,b,c]

1. Introduction

The development of the smart grid has been driven by rising concerns about global warming and the requirement to lower power costs. The outdated power grid is widely acknowledged to be inefficient, unreliable, polluting, unidirectional (from producer to consumer), uses bulk production, has few sensors, and provides only a limited amount of assistance for automation control and operations. A smart grid, on the other hand, is digital, incorporates distributed energy resources, is automated, allows for two-way communication, and incorporates cutting-edge ICT infrastructure, including wireless sensor networks, cloud computing, and edge/fog computing. Demand response management, advanced metering infrastructure, substation automation, home energy management systems, outage management, distributed automation, asset management, electric vehicle charging, distributed energy resources and storage, and wide-area situational awareness systems are just a few of the advanced services that were made possible by the smart grid (Hamidieh and Ghassemi, 2022).

[a] Department of Computer Science, Hekma School of Engineering, Computing and Design, Dar Al-Hekma University, Saudi Arabia.
[b] Department of Computing, University of Turku, Finland.
[c] Department of Technology, Higher Institute of Computer Sciences and Mathematics, University of Monastir, Tunisia.
Email: imed.bendhaou@utu.fi

For decades, the microgrid has been an effective way to provide power to remote communities. Manhattan Pearl Street Station, which was built to produce 1,100 kW DC using a steam engine, was the first microgrid to be publicly acknowledged (Cunningham and Paserba, 2022). In the last decade, microgirds have proven to be a viable solution to accommodate the ever-increasing demands on Distributed Energy Resources (DERs).

There are two primary types of microgrids. The initial type is called an AC microgrid, which utilizes an AC bus to supply loads with AC voltage. The second type is known as a DC microgrid, where loads are connected to DC electricity through a DC bus. Figure 1 illustrates these two distinct forms of microgrid. Furthermore, it should be noted that hybrid microgrids are also a viable option.

A DC microgrid is a localized power system that operates using direct current (DC) voltage. It typically includes multiple energy sources such as solar panels, wind turbines, and batteries, as well as energy users such as homes and businesses. DC microgrids provide a reliable and efficient power supply for small communities or individual buildings that can operate independently or in combination with the main power grid. They are also ideal for areas with limited access to traditional power sources, such as remote rural communities or disaster-stricken areas.

In recent years, DC microgrids have gained popularity due to their ability to improve the resilience of the general power grid and provide a pathway to the integration of renewable energy. It is well accepted that the DC microgrid has better efficiency and provides better power quality compared to the AC microgrid (Lotfi and Khodaei, 2017; Rangarajan et al., 2023). Table 2 provides a comparison between the AC and DC microgrids. (Lotfi and Khodaei, 2017) found that DC microgrids are the most cost-effective option for microgrids with a high proportion of DC loads.

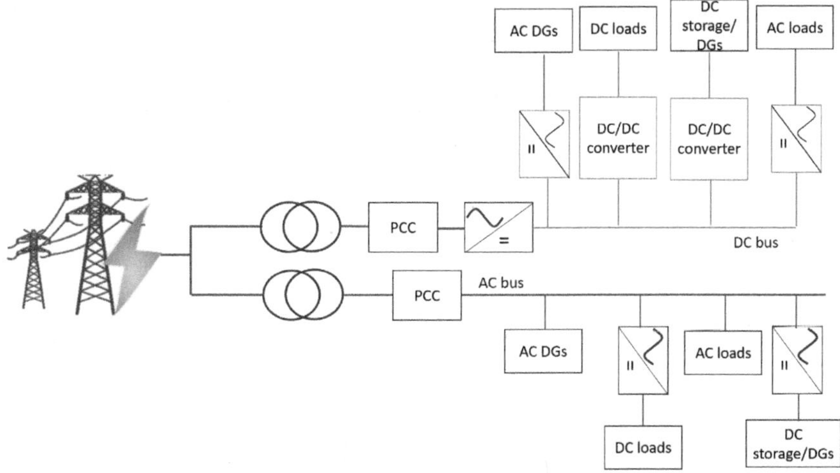

Figure 1. AC and DC microgrids connected to the grid with distributed generations.

Table 1. Comparison between AC and DC microgrids.

Feature	AC Microgrid	DC Microgrid
Power type	Alternating Current (AC)	Direct Current (DC)
Voltage	Single-phase or three-phase, 120V/240V or 277V/480V	Typically 380V
Power conversion	Not required for AC loads	Required for AC loads
Efficiency	Lower efficiency due to line losses	Higher efficiency
Stability	Less stable due to external disturbances	More stable
Cost	Lower initial cost	Higher initial cost
Maturity	More mature y	Less mature

Over the years, companies started offering DC appliances. As reported in (Sabry et al., 2020), DC appliances have a better efficiency than their AC counterparts. Table 2 samples the characteristics of some DC appliances found in today's market.

Table 2. DC appliances.

Appliance	Rated Power	Rated Voltage
Air conditioning (12,000 BTU)	Heating 1050 W Cooling 980 W	48V
Washing machine (7KG)	Washer power - 300W Dryer power 140 W	12V
Microwave (12L)	800W	24V
Induction cooker	1200 W	48V
All-in-One Desktop	90W	20V

Furthermore, the growing availability and proliferation of information and communication technology (smartphones, Internet of Things, printers, scanners, etc.) has favoured the adoption of DC microgrids. Communication networks, personal computers, and data centers are the three main types of ICT equipment. (Van Heddeghem et al., 2014) predicts that the power consumption of the three groups will double every 10 years. ICT equipment consumed about 330 TWh in 2012. This trend requires rapid action to reduce the carbon footprint of ICT equipment.

In the wireless communication domain, 77% of the energy consumed by ICT is attributed to the network operator (Van Heddeghem et al., 2014). As shown in Figure 2, a base transceiver station (BTS) is composed of a power unit, a cooling and lighting system, a Radio Unit (RU), a Base Band Unit (BBU), an Active Antenna Unit (AAU), and a backhaul network. The RU consumes more than 60% of the overall power, followed by the cooling system, which consumes 25%. (Tradacete et al., 2021) proposed a microgrid with an energy management system that eliminates noncritical loads based on meteorological conditions (temperature, humidity, irradiance, wind speed and wind direction), the status of the battery storage energy system, and the price of the energy to make the BTS more environmentally friendly.

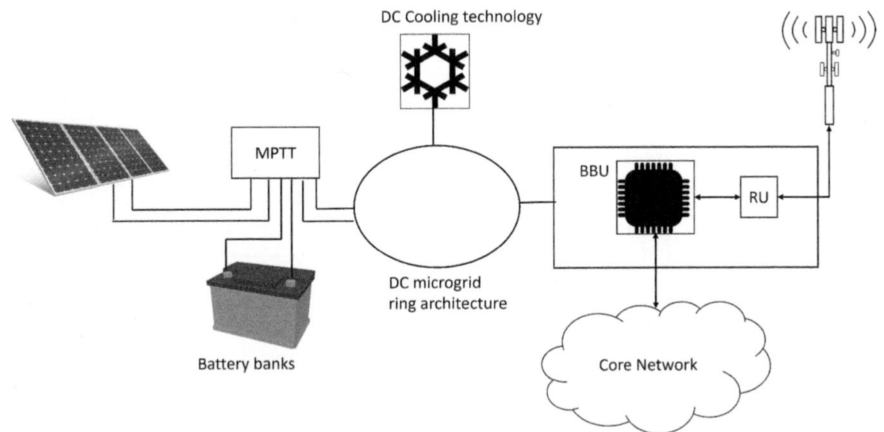

Figure 2. Block diagram of a green base transceiver station.

The following sections of this chapter are structured as follows. In Section 2, we investigate the typical architectures of DC microgrids. Section 3 looks at the essential power electronic components used in the design and implementation of DC-microgrids. Section 4 explains the control of power electronics devices with the help of information and communication technologies. Finally, Section 5 summarizes the chapter.

2. Architecture of DC microgrid

There are many types of microgrid architecture: single-bus, multi-bus, ring-bus, zonal. multiterminal and partially grid-connected microgrid (Dragičević et al., 2016) (Shaban et al., 2021).

2.1. Single-bus DC microgrid

Is a widely used architecture as it is low-cost. In this architecture, loads, ESSs, and DERs are all connected to a single bus. The single-bus DC microgrid comes in two categories: unipolar and bipolar. The advantages of the single-bus DC microgrid lie in the simplicity of the design of control algorithms, which are very stable and reliable. The bipolar single-bus architecture has been developed to increase the immunity of the power system from failure. A laboratory-scale prototype of a bipolar DC microgrid has been reported in (Kakigano et al., 2010). The system consists of a 340V VDC. Three voltage levels are used in the microgrid: +170 V line, ground and the –170V line, which gives the DC-DC converter the freedom to choose 340V, 170V, or –170V as input voltage. This kind of redundancy guarantees power availability for critical loads.

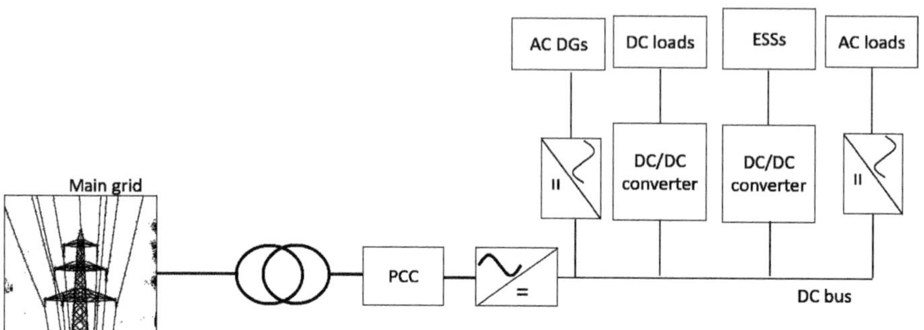

Figure 3. Toploggy of a single bus DC microgrid.

2.2. Multi-bus DC microgrid

It is a topology that has been advocated to increase the flexibility of the DC microgrid. The chief idea is to offer loads of options to choose power from multiple sources. There are several DC buses in the microgrid, and each bus has a number of DERs, loads, and ESSs. Isolated DC-DC converters link these buses together. Figure 4 illustrates the simplified architecture of the multi-bus DC system. The system is highly reconfigurable and more reliable than a single-bus DC microgrid, since each load can get power from several sources that are located on the same or distinct DC buses. The multi-bus DC microgrid's control system is more intricate than the single-bus system. The three control layers should regulate the voltage and balance the powers between the buses (Ballal et al., 2022).

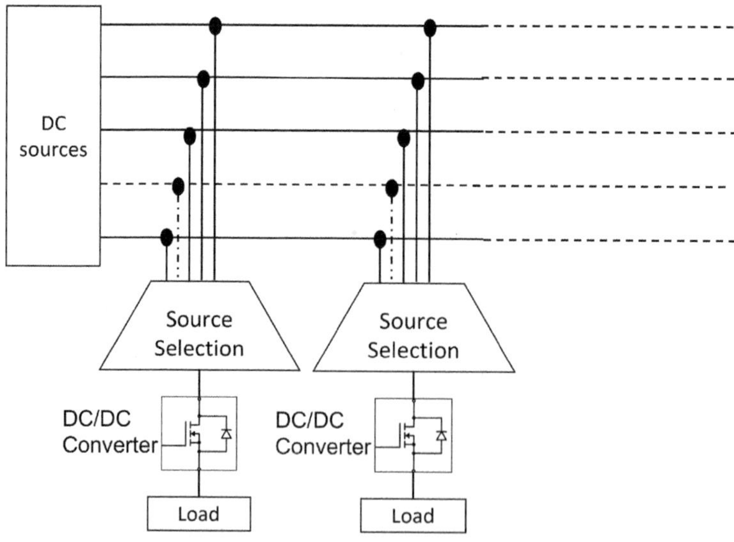

Figure 4. Topology of a multi-bus DC microgrid.

2.3. Ring bus DC microgrid

A reconfigurable DC microgrid can be achieved using a number of topologies among them is the ring-bus architecture. As illustrated in Figure 5, the bus consists of DERS, loads, and storage units connected by a ring-dc bus. Each module is connected through a proper AC-DC or DC-DC converter. These converters are interfaced to the bus through an Intelligent Electronic Breaker (IEB).

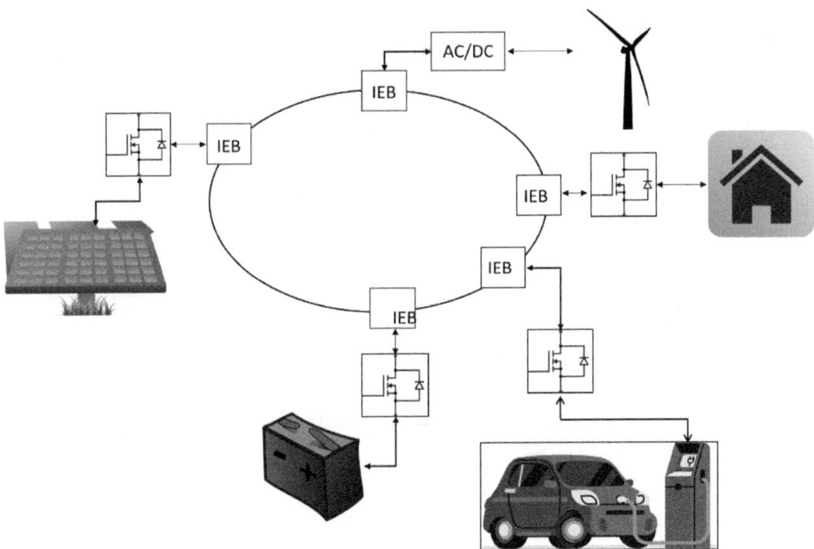

Figure 5. Topology of a ring-bus DC microgrid.

The salient features of the ring-bus architecture are its scalability, reliability, and efficiencies, among other things. However, compared to radial and mesh topologies, the ring-bus DC microgrid requires an elaborate scheme for the localization and isolation of faults. An Intelligent Electronic Device (IED)-based scheme has been devised by (Park et al., 2013). To isolate faults, the overcurrent and differential current of neighboring IEDs. The circuit breaker is set to open or close depending on the nature of the current fault (permanent or cleared). Intelligent electronic devices (IEDs) are embedded systems enabled by IoT that are used to protect, control, and monitor power systems. The architecture of a typical IED consists of a processor, a main memory, IOs, and a communication interface. The IDEs devised in (Park et al., 2013) are prototyped using a 32-bit microcontroller unit (MCU) clocked at 150 MHz.

2.4. Partially gird-connected DC-microgrid architecture

The intermittent nature of the PV system along with the asymmetric power demands calls for effective power management and the implementation of a de-

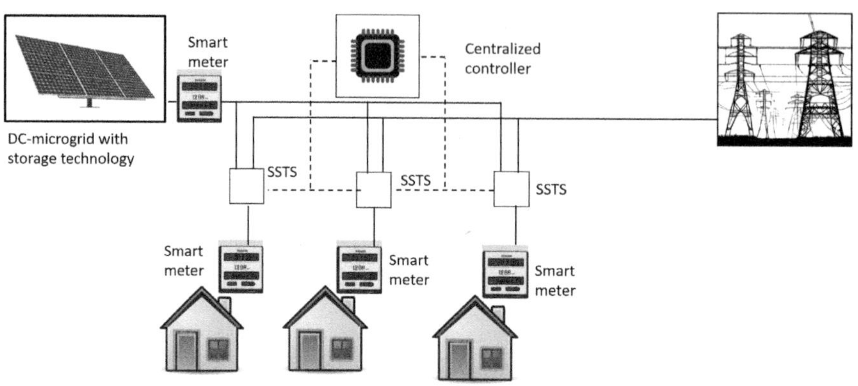

Figure 6. Topology of a partially grid-connected microgrid.

mand response program. In this context, (Shaban et al., 2021) devised a microgrid architecture and control algorithm that connects a suitable number of loads to the main grid in case the power demands are higher than the power generated by renewables. The architecture shown in Figure 6 is composed of the following elements: a DC microgrid with an energy storage system (ESS), solid state transfer switches, smart meters, and a centralized controller.

3. Power electronics for the DC microgrid

Power electronics are a crucial enabler technology for the development and deployment of DC microgrids. As discussed in the previous section, the DC microgrid uses a multitude of power electronic converters and energy storage systems. The typical power electronic devices used in the DC microgrid are summarized in the Table 3. This section focuses on describing the architecture of typical DC-DC converters used in the design of a low-voltage DC Microgrid (LVDCMG).

3.1. DC-DC converters

DC-DC converters are power electronic devices that convert DC voltage from one level to another. Step-up DC converters are used to increase the input voltage (Forouzesh et al., 2017). On the contrary, the output voltage of the DC-DC step-down converters is lower than the input voltage. Typically, step-down DC converters are used to power DC appliances from a high-voltage DC bus, whereas step-up DC-DC converters are used to connect DC sources to the DC bus.

There exist numerous taxonomies for DC-DC converters. These classifications take into account the topology, control techniques, operation mode, conversion ratio, isolation (Kasper et al., 2014; Páez et al., 2019; Salem et al., 2018).

Table 3. Power electronic components.

Component name	Explanation
Buck Converter	Converts DC voltage from high value to lower value
Metal-Oxide-Semiconductor Field-Effect Transistors (MOSFET)	Amplification or switching
Insulated Gate Bipolar Transistors (IGBT)	Control and speed regulation
Voltage regulators	Maintain a constant output voltage
Inductor	Stores magnetic energy (current source)
Capacitor	Stories electric energy (voltage source)
Battery	Stores energy
Photovoltaic	Converts sunlight source to electricity
Maximum Point Power Tracking (MPPT) controller	A device that extracts the maximum power from a solar panel
onverter	A DC-DC converter that amplifies an input voltage
Buck-Boost Converter	A DC-DC converter that either amplifies or reduces the magnitude of the input voltage
Fly-back DC-DC converter	An isolated Buck-Boost converter

The input and output circuits of isolated DC-DC converters are electrically separated from each other by a transformer. Isolation is designed to protect the load from electric shock and reduce noise and interference. Isolated DC-DC converters are used in the DC microgrid to convert energy from renewable sources, such as wind turbines and solar panels.

Nonisolated DC-DC converters, on the other hand, do not isolate the input and output. These converters are best suited for low-power applications. They are particularly useful for step-down voltage conversion. In the electric vehicle domain, the charging station may make use of the nonisolated bidirectional DC-DC converter.

Typical equations used to characterize any given converter are summarized in Table 4. Details of those equations are reported in (Kazimierczuk, 2016).

3.1.1. DC-DC Boost converters

The step-up converter converts a low input voltage to a high output voltage by storing the input voltage in an electric storage element (inductor or capacitor) and then releasing it at the output.

There exist two types of boost converters: isolated and non-isolated. Isolated converters use a transformer to isolate the input and output. It is usually used to power sensitive appliances and to connect DC microgrids (Lee et al., 2015). In the design of a microgrid, an isolated bidirectional DC-DC converter is used to connect the energy storage system to the DC bus. Figure 7 illustrates a current-fed isolated bidirectional DC-DC converter devised by (Choi et al., 2022). The converter outperforms the voltage-fed dual active bridge converter, as it has better efficiency under heavy load variation and guarantees a zero voltage switching condition.

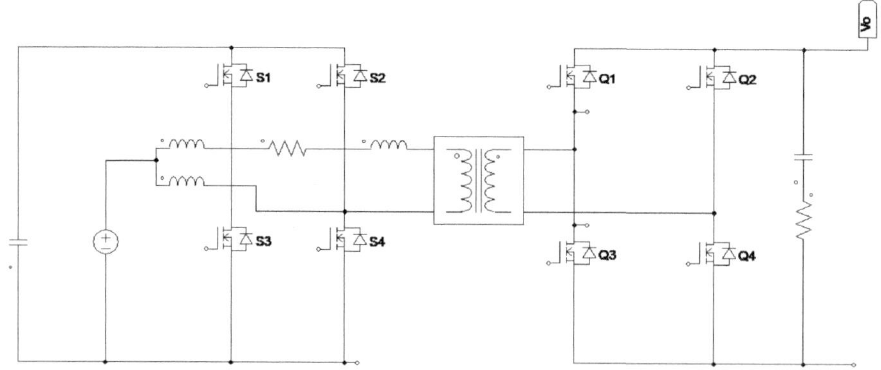

Figure 7. Isolated bidirectional DC-DC converter proposed by (Choi et al., 2022).

Table 4. Characterization of DC-DC converters.

Equation	Explanation
$P_{in} = V_{in} I_{in}$	Input power in Watt.
$P_{out} = V_{out} I_{out}$	Output power in Watt.
$\eta = \frac{P_{out}}{P_{in}}$	Efficiency of the converter.
$P_{loss} = P_{in} - P_{out}$	Power loss in Watt.
$H_V = \frac{V_{out}}{V_{in}}$	DC voltage transfer function.
$H_I = \frac{I_{out}}{I_{in}}$	DC current transfer function.
$LNR = \frac{\Delta V_{out}}{\Delta V_{in}}$	Line regulation in mV/V at a constant temperature and constant I_{out}.
$LOR = \frac{\Delta V_{out}}{\Delta I_{out}}$	Load regulation in mA/V at a constant input voltage and constant temperature.
$PLOR = 100\% \frac{VOC - V_{out,full}}{V_{out,full}}$	Percentage load regulation when input voltage and temperature are constant.
$LLR = 100\% \frac{\frac{\Delta V_{out}}{V_{out,nom}}}{\Delta I_{out}}$	Line/load regulation in $\frac{\%}{A}$ at a constant input voltage and constant temperature.
$R_{out} = - \frac{\Delta V_{out}}{\Delta I_{out}}$	Dynamic output resistance when both the temperature and input voltage are held constant.
$THR = 100\% \frac{\Delta V_{out}}{\Delta P_{diss}}$	Thermal regulation in $\frac{\%}{W}$ computed at a constant V_{in} and I_{out}.
$RRR = \frac{V_{IR}}{V_{OR}}$	Computed Ripple Rejection Ratio (in dB) given the input ripple (V_{IR}) and the output ripple (V_{OR}).
$f_0 = \frac{1}{2\pi \sqrt{LC}}$	Corner/ cutoff frequency (in Hz) of an LC low-pass filter.

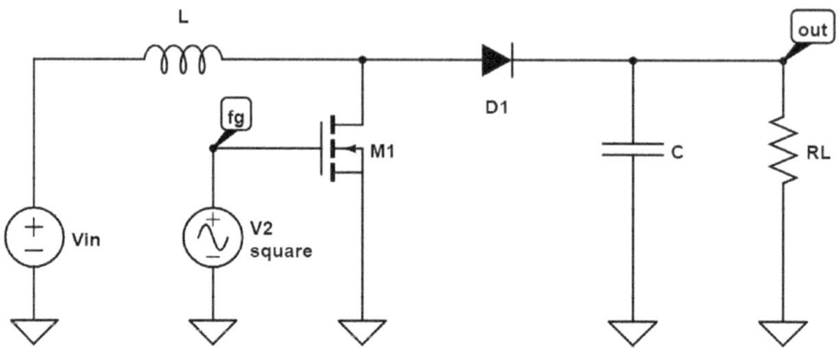

Figure 8. Circuit diagram of a PWM boost converter.

A PWM DC-DC boost converter is widely used to amplify the input voltage. The converter consists of an inductor, a switch, a diode, a control block, and a capacitor. The metal oxide field effect transistor (MOSFET) is commonly used as a switch. The operation of the converter is arbitrated by pulse width modulation (PWM). Figure 8 shows the PWM boost converter.

The boost converter works in a Continuous Conduction Mode (CCM) or a Discontinuous Conduction Mode (DCM). The mode of operation depends on the charging and discharging rate of the inductor.

In this section, the operation of the boost converter in the CCM mode is considered. For the DCM mode, the reader can refer to (Kazimierczuk, 2016). In the ideal case, the peak-to-peak value of the ripple current at the inductor is computed using

$$\Delta I_L = \frac{V_{out}\lambda(1-\lambda)}{FL},\tag{1}$$

where λ is the duty cycle determined using (2), F is the frequency of the PWM, and L is the value of the inductor.

$$\lambda = 1 - \frac{V_{in}}{V_{out}}.\tag{2}$$

$$L_{min} = R_L D\frac{(1-D)^2}{2F},\tag{3}$$

where $R_L = \frac{V_{in}^2}{P}$ is the load resistance computed, and P is the input power of the converter. The capacitance is determined using

$$C_{min} = \frac{\lambda}{R_L \times F \left(\frac{\Delta V_{out}}{V_{out}} \right)}. \tag{4}$$

Example 3.1. Let us consider the case of designing a DC boost converter with the following parameters: $V_{out} = 24\text{V}$, $V_{in} = 12\text{V}$, $F = 40\text{kHz}$, $\frac{\Delta V_{out}}{V_{out}} <5\%$, $P = 100$ W. The converter should operate in the CCM mode.

The duty cycle $\lambda = 1 - \frac{12}{24} = 50\%$. The minimum value of the inductor is $L_{min} = 9\mu\text{H}$. To operate in CCM mode, we increased the value of the inductor by 25%, that is, L$\approx$ 586. The value of the load resistance $R_L = \frac{24^2}{100} = 5.76\Omega$. The capacitance is $C_{min} = \frac{0.5}{5.76 \times 0.05 \times 10^3} \approx 43.4\mu\text{F}$. The output voltage and the inductor current are shown in Figure 9. The output voltage settles after 882 μ sec and oscillates between 24.07V and 22.9V. Figure 9 shows the simulated boost converter.

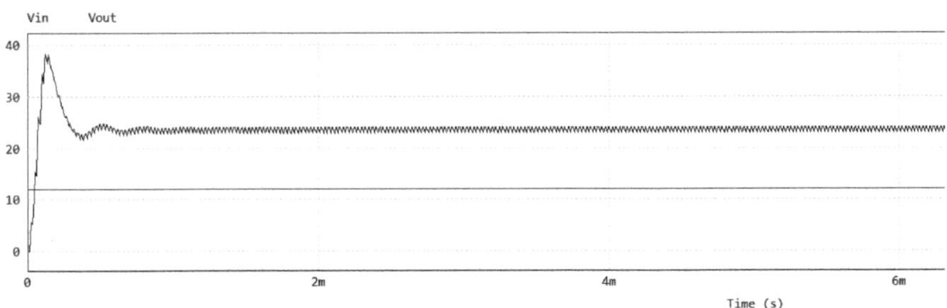

Figure 9. Output voltage of a simulated boost converter, where $V_{in} = 12\text{V}$ and $V_{out} = 24\text{V}$.

3.1.2. PWM Buck Converter

A pulse-width dc-dc buck (chopper) converter is a step-down converter which is composed of switches, an inductor, a capacitor and a load. A typical Buck converter circuit topology is schematically shown in Figure 10.

The converter can operate in continuous or discontinuous conduction mode. The value of the current stored in the inductors distinguishes the boundary between the CCM and DCM modes. The inductor current does not reach zero in the CCM mode. In DCM, the inductor current reaches zero and remains there for a period of time before increasing. The crucial mode (CRM) is a state that exists between the CCM and DCM mode.

Assuming the ideal diode and transistor, in the CCM mode, the ripple current of the inductor, the output voltages, and the minimum capacitance value are given

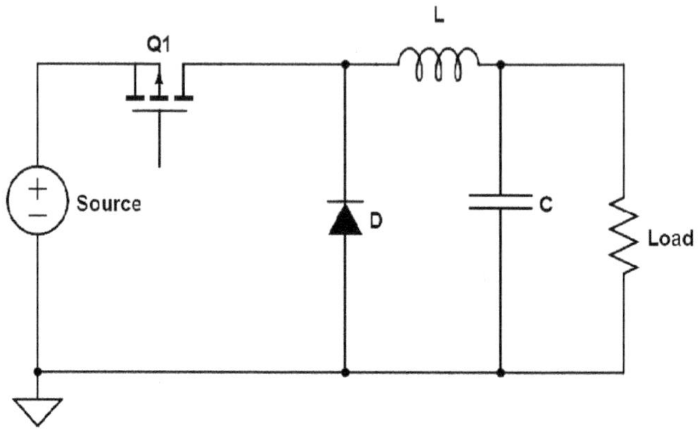

Figure 10. DC-DC Buck converter.

by (5) and (6), respectively.

$$\Delta I_L = \frac{V_{out}(1 - \lambda)}{FL},$$

(5)

where L is the inductor, λ is the duty cycle, and F is the frequency of the PWM. The minimum value of the inductor, L_{min} is computed using (6)

$$V_{out} = \lambda V_{in}.$$

(6)

$$C_{min} = \frac{1 - \lambda}{8LF^2 \left(\frac{\Delta V_{out}}{V_{out}} \right)}.$$

(7)

The minimum value of the inductance, L_{min} is determined using (8).

$$L_{min} = R_L \frac{1 - \lambda}{2F}$$

(8)

Example 3.2. Let us consider the case of designing a DC buck converter with the following parameters: $V_{out} = 12\text{V}$. $V_{in} = 48\text{V}$, $R_L = 50\Omega$, $F = 40\text{kHz}$, $\frac{\Delta V_{out}}{V_{out}} < 5\%$. The converter should operate in CCM mode.

The duty cycle $\lambda = \frac{12}{48} = 25\%$. The minimum value of the inductor is $L_{min} = 50\frac{0.75}{80 \times 10^3} = 468.75\text{uH}$. To operate in CCM mode, we increased the value of the inductor by 25%, that is, L$\approx$ 586. The capacitance value is $C_{min} = \frac{0.75}{8 \times 586^{-6}0.05(40 \times 10^3)^2} \approx 2$ uF. The output voltage and inductor current are shown in Figure 11. The output voltage settles after 510 μ sec and oscillates between 12.24V and 11.65V.

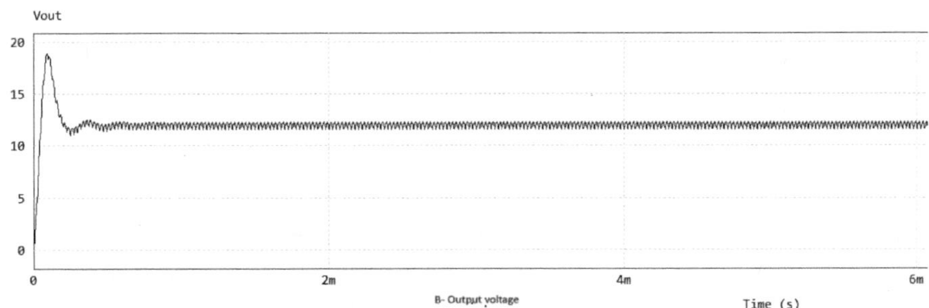

Figure 11. Simulation results of the buck converter, where $V_{in} = 48V$ and $V_{out} = 12V$.

3.2. *Battery storage system*

The intermittent nature of solar energy can be solved using Energy Storage Systems (ESSs). The various technologies used to store energy are presented in Table 5 (Kim and Chou, 2015; Zhang et al., 2021). Batteries are among the most widely used energy storage technologies in the DC microgrid. Two types of competing battery technologies are available on the market: Lithium-ion and lead acid batteries. The features of each type of battery are reported in Table 6.

3.3. *Electric equivalent circuit models of batteries*

Numerous techniques have been devised to estimate the SOC of batteries. The most popular are the look-up table (LUT)(Einhorn et al., 2013), open circuit voltage (OCV) (Weng et al., 2014), Kalman filter (Hu et al., 2012; Li and Choe, 2013), and ECM (equivalent Circuit Model). The comparison between those techniques is summarized in Table 7.

A hybrid method that combines more than one technique has improved the accuracy of SoC estimation with reduced complexity. The authors in (Misyris et al., 2019) reduced the SOC errors under the Dynamic Stress Test (DST) by using a hybrid coulomb counting and adaptive filtering technique based on the Extended Kalman Filter (EKP).

The equivalent circuit model is used for the estimation of SOC. The simplest model for the battery is to approximate the battery by its equivalent Thevenin circuit considering only its internal resistance, which is inaccurate in estimating SoC. An improved version has been developed, which is depicted in Figure 12 (He et al., 2011). The circuit consists of the ohmic resistance of the battery (R_O) in series with a parallel RC network(R_p, C_{tr}). Both the polarization resistance (R_p) and the transient capacitance (C_{tr}) is determined using the characteristics of the battery. The output voltage given the load current I_L is computed using (9).

$$V_L = V_{OC} - (V_{TH} + R_O I_L), \tag{9}$$

Table 5. Energy storage system.

Name	type	characteristics
Flywheel energy storage	mechanical	has a low maintenance cost and high power density
Compressed air energy storage	mechanical	stores energy for a long time and has an efficiency close to 80%
Hydrogen storage	Chemical	has a high-density power, clean, but it has issues with safety and efficiency
Battery	Electrochemical	provides consistent voltage output and has a low self-discharging rate, but it takes a long time to charge
Supercapacitor	Electrochemical	has a high cycle-life, and fast charge time, but suffers from low energy density

Table 6.   Comparison between lead acid and Lithium-Ion batteries.

Parameter	Lithium-Ion	Lead acid
Capacity, Q	100–200 Wh/kg	30–50 Wh/Kg
Internal resistance, R_{in}	low	high
Operating temperature range, T	from –20 to 60 °C	from –10 to 50 °C
Cycle life	1000-5000 cycles	300–500 cycles
Safety	High	Low
Estimated cost (USD/ KWh)	209	186
Self-discharge rate per month	2–3%	5–10%

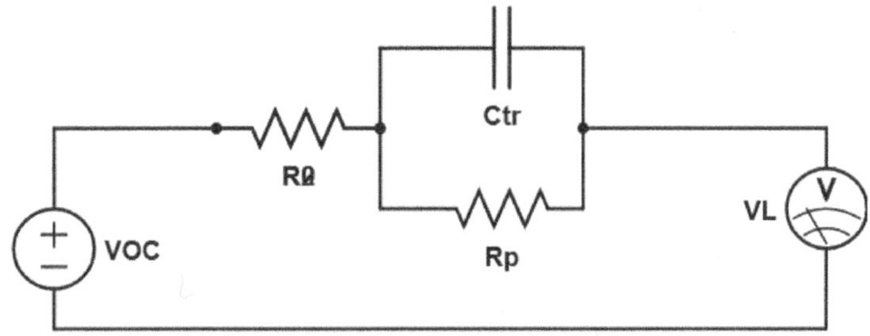

Figure 12.   Thevenin ECM for battery.

where V_{OC} is the open circuit voltage of the battery, V_L is the load voltage and V_{th} is voltage across the capacitance C_{tr}. The open circuit voltage is dynamic and accounts for the temperature, SOC and SOH of the battery.

A dual polarization model with a better dynamic performance compared to the Thevenin model is shown in Figure 13. The output voltage given the current load (I_L) is calculated using (10)

$$V_L = V_{OC} - (I_L R_O + V_{PA} + V_{PC}), \tag{10}$$

where V_{PA} and V_{PC} are, respectively, the voltage across the capacitance C_{trA} and C_{trc}.

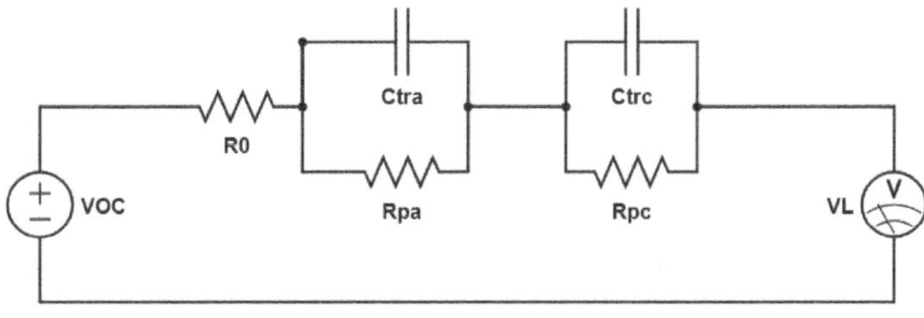

Figure 13.   ECM model based on dual polarisation.

Table 7. Comparison of battery modelling methodologies.

Method	Pros	Cons
Look-up table	Simple and straightforward	Sensitive to changes in operational conditions, such as age
Coulomb counting	Simple to implement	Sensitive to changes in operational conditions, such as age
Open circuit voltage	Very accurate	The battery needs to rest for an extended period of time
Kalman filter	Resilient to noise and modelling errors Reliable and suitable for online SOC estimation	Computationally appealing
Artificial Intelligence	Able to learn the complex relationship between battery perimeters	Prone to underfit and overfit
Equivalent circuit model	Simple calculation Has a clear physical meaning	Parameters need to be dynamic

4. Control and communication for the DC microgrid

The control of DC microgrids is a challenging task due to the uneven distribution of energy resources and the dynamic changes in energy demand. One of the main challenges in the operation of DC microgrids is distributing the power load evenly among distributed energy resources (DERs) and energy storage systems (ESSs). Additionally, the clustering of DC microrgirds contributed to increasing the complexity of the control techniques.

To address control challenges, hierarchical control of DC microgrids has been proposed in many published reports (Dragičević et al., 2018; Gao et al., 2018; Guerrero et al., 2009; OK̃eeffe et al., 2017). Control of DC microgrids consists of three layers: primary control, secondary control, and tertiary control. The summaries of the primary objective of each control level are depicted in Figure 14.

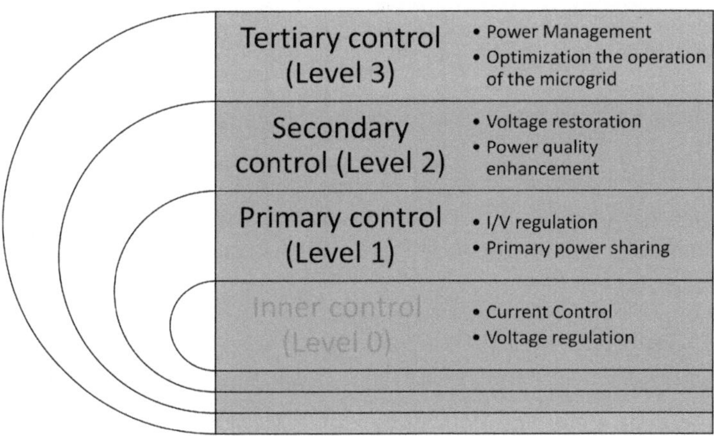

Figure 14. Hierarchical control of DC microgrids.

Voltage droop control has been devised as an effective communicationless and decentralized technique for sharing load power. It regulates the output of DC converters.

4.1. Droop control

Droop control is a technique used in DC-DC converters to regulate voltage output when multiple converter modules are connected in parallel. Its purpose is to ensure an even distribution of the load among these parallel-connected modules, which collectively power a common load.

In the droop control method, a small portion of the output voltage generated by each converter module is routed back to a shared line, forming a control signal. Subsequently, this signal is utilized to manage the output voltage of each module. The goal is to ensure that each module responds uniformly to variations in load current and voltage.

As the load current increases, the output voltage of each module decreases proportionally according to the droop resistance. This enables each module to contribute its fair share of the increasing current while maintaining a balanced output voltage alongside the other modules connected in parallel. In the sequel, we will initially delve into the fundamentals of droop controllers and then explore suggested approaches for enhancing the stability of DC microgrids through the use of droop control.

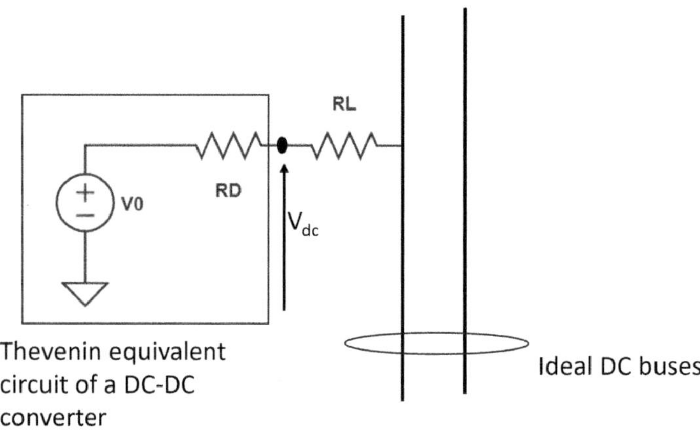

Figure 15. Circuit model of a DC microgrid with one DC-DC converter.

Given the equivalent Thevenin circuit of a DC-DC converter, as shown in Figure 15, the voltage output of the converter is determined using (11).

$$V_{DC} = V_O - R_D I_{DC}, \qquad (11)$$

where V_{DC} is the voltage output of the converter, V_0 is the Thevenin output voltage, R_D is the droop gain (virtual resistance), and I_{DC} is the converter output current.

The power load sharing among N parallel converters is achieved by using virtual resistance. Figure 16 illustrates the characteristics of droop of V-I.

Let VDC_N be the nominal voltage of the DC bus, I_{max} be the maximum output current, and ΔV the maximum voltage deviation allowed. The upper limit for the droop gain (R_D) is determined using (12) and the output voltage of the converter is computed using (13) (Guerrero et al., 2009).

$$R_D = \frac{\Delta V}{2I_{max}}. \qquad (12)$$

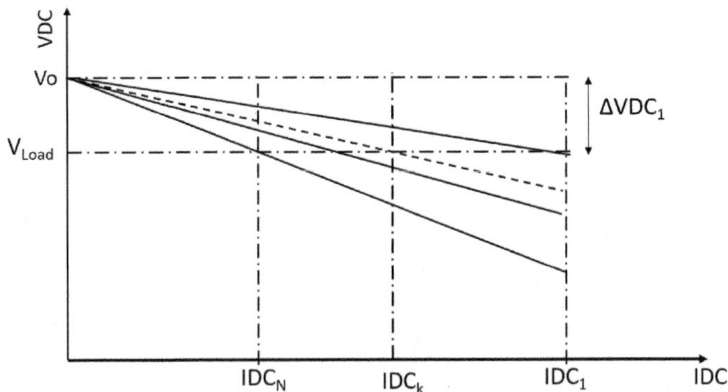

Figure 16. Droop curves for N parallel converters.

$$V_0 = VDC_N - \frac{\Delta V}{2}.$$ (13)

The control of the converters is done by regulating the duty cycle of the pulse-width modulator (PWM). Figure 17 shows the droop control scheme which is made up of a compensator, voltage and current sensors and a PWM generator. This scheme was devised by (Lu et al., 2014).

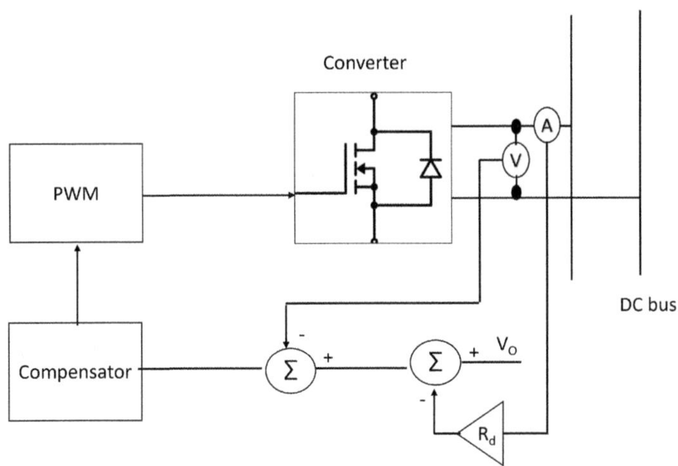

Figure 17. Primary control using V-I droop controller.

4.2. Secondary control

The secondary control of the DC microgrid attempts to maintain the voltage and current levels at their designated/nominal values. However, droop control is not successful in evenly distributing power among multiple converters due to discrepancies in voltage and inaccuracies in current and voltage measurements.

For the voltage droop control, the secondary control measures the voltage level at regular intervals at the DC bus and compares it with the nominal value using a voltage comparator. Let K_p and K_{int} be the parameters of the secondary control compensator; the droop shift value is calculated using (14).

$$\delta v_o = K_p \times (V_O - \overline{V_{DC}}) + K_{int} \int (V_O - \overline{V_{DC}})dt, \tag{14}$$

where $\overline{V_{DC}}$ is the average measured voltage at the DC bus.

The voltage correction term, δv_o, is then communicated through the droop controller at the primary control level. As pointed out by (Gao et al., 2018), the secondary controller can be realized using the centralized (supervisory), decentralized, or distributed method.

In the centralized approach, a centralized proportional integral, PI, controller collects through a low-data rate communication protocol sensed parameters and sends the droop shifting value (δv_o) to the primary controllers. The method is vulnerable to the single point of failure (SOPF).

A distributed architecture has been designed to address the issues related to the supervisory control approach, which involves the installation of a secondary controller in each converter. This architecture requires multiple communication links between adjacent converters, as illustrated in Figure 18.

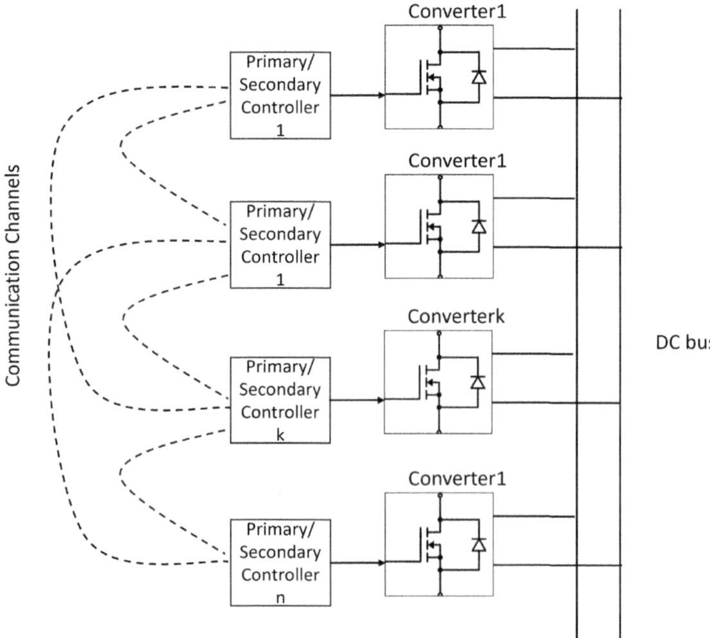

Figure 18. Distributed secondary level control for DC microgrid.

The distributed secondary control approach induces a very complicated communication network. Given n, the number of converters, the complexity of the communication link is complex $\mathcal{O}(n^2)$. To overcome this limitation, three classes of decentralized control have been devised: the first uses power line communication (Pinomaa et al., 2011), the second is based on DC bus signalling (Zhang et al., 2011), and the third uses adaptive droop calculation method (Augustine et al., 2015).

4.3. Tertiary control

The tertiary control of a DC microgrid is the highest level of control techniques, with the primary objective of optimizing the operation of the microgrid and reducing operational costs through the implementation of advanced power and energy management algorithms such as genetic algorithms, particle swarm optimization, consensus, and machine learning algorithms. As reported by (Abhishek et al., 2020; Bidram and Davoudi, 2012; Guerrero et al., 2009), the services provided by tertiary control include, but are not limited to, the cost-effective operation of the microgrid, the management of the power flow between microgrid clusters and the utility grid, and the sharing of power between multiple DC microgrids. This type of control is implemented in two ways: centralized and distributed. It serves as a bridge between the business and the secondary control layer (Wu et al., 2022).

The centralized tertiary controller monitors the current in the microgrid by using a static bypass switch and compares it to a desired current value. If the desired current is negative, the controller will inject power from the main grid (Guerrero et al., 2009). The multilevel energy management system (EMS) (Xiao et al., 2016) has a tertiary controller that performs economic dispatch by comparing the marginal costs of the system units. This is done to reduce the cost of system operation. The power references of the system units are determined centrally, with the unit having the lowest marginal cost being given a higher priority for utilization, and the opposite for the one with the higher marginal cost. EMS uses a Supervisory Control and Data Acquisition (SCADA) system for centralized coordination. This SCADA system has been designed with a programmable logic controller (PLC) as the central controller, and the communication link is based on the standard Modbus TCP/IP protocol. In the system designed by (Shaban et al., 2021), the tertiary control receives information on power consumption in real time via the Lora communication protocol. If demand exceeds the energy produced by renewable sources, the consumer with the highest demand is reconnected to the primary grid.

The centralized tertiary controller has a single point of failure and is not effective in regulating power flow when the voltage deviation is significant. To address these problems, a distributed architecture has been proposed in (Moayedi and Davoudi, 2016) for DC microgrid clusters. The cooperative controller, based on a multi-agent system, adjusts the voltage set points for each microgrid using a distributed ternary controller. A novel distributed architecture in (Mudaliyar et al., 2020) has been proposed in (Mudaliyar et al., 2020) to manage the power flow in DC microgrid clusters (DCMGCs). This approach combines the cost of operation of distributed generators with the voltage regulation of DCMGCs.

4.4. Discussion

Communication technologies play a pivotal role in shaping the secondary and tertiary control aspects of the DC microgrid. Within the existing literature, various communication protocols stand out for designing DC microgrid controllers, including fiber-optic, power-line communication (PLC), controller area network (CAN), Ethernet (such as IEC 61850), wireless technology, and cellular technology (spanning 2G, 3G, 4G, and 5G).

IoT has emergeed as a viable solution for smart-grid due to its attributes of scalability, efficiency, reliability, and cost-effectiveness (Ghasempour, 2019) (Dhaou et al., 2022). It has ushered in the transition from a conventional microgrid to an intelligent, smart microgrid, seamlessly integrating advanced technologies like smart sensors, edge/fog computing, and cloud servers. The incorporation of these cutting-edge communication and sensing technologies has facilitated the implementation of robust control and monitoring algorithms, leveraging machine learning algorithms, Distributed Ledger Technology, and blockchain technology.

In particular, IoT-based control schemes address the limitations associated with centralized control, offering decentralized control mechanisms that enhance the autonomy of agents operating within the microgrid. Moreover, IoT networks contribute to precise power generation and demand forecasting, thereby significantly improving the overall performance and reliability of the microgrid. This transformative integration of IoT technologies represents a paradigm shift in the efficiency and capabilities of DC microgrid systems (Ansari et al., 2021).

5. Summary

In the past decade, there has been a great deal of advancement in photovoltaic system technology, such as crystalline silicon, thin-film, perovskite solar cells, and organic photovoltaics. Furthermore, energy storage technologies, semiconductor power electronics, advanced control systems, architecture, and reliable communication protocols have also seen a rapid development. These and other elements have made DC microgrids more popular in both academic and corporate settings. DC microgrids have garnered increased interest due to their notable attributes of cost-effectiveness, dependability, and efficiency when juxtaposed with traditional AC microgrids. In this chapter, a comprehensive exploration of DC microgrid architectures and technologies, with a specific focus on the topologies of DC microgrids, DC-DC converters, control mechanisms, and storage technologies has been reported. The selection between DC or AC-microgrids is contingent upon the number of AC appliances and the specific type of distributed renewable energy resources.

Careful planning and meticulous selection of technologies and architecture are paramount for successful DC microgrid implementation. This necessitates considerations for renewable energy resources, DC-DC converters, inverters, energy storage systems, as well as communication and control units. Among the various DC microgrid topologies, single bus, ring bus, multibus, and partially grid-connected configurations are identified as the four most common.

In the storage domain, the lithium-ion battery emerges as the predominant technology for energy storage systems. The sizing of mangement of battery system is tied to energy demands and generation. This chapter delves into battery modeling, placing a particular emphasis on the electric circuit model. The combination of the dual-polarity model and Thevenin's equivalent circuit model, is deemed as a means to accurately determine the battery's state of charge.

The management of a microgrid and microgrid cluster is achieved through a three-tier control system, which is explored in depth in this chapter. Communication assumes a critical role, particularly in secondary and tertiary control, with both communication technologies and security playing pivotal roles in ensuring grid stability. Subsequent chapters will delve deeper into communication and security issues, additionally exploring the utilization of machine learning and artificial intelligence in battery state estimation.

References

Abhishek, A., Ranjan, A., Devassy, S., Kumar Verma, B., Ram, S. K. and Dhakar, A. K. 2020. Review of hierarchical control strategies for dc microgrid. IET Renewable Power Generation, 14(10): 1631–1640.

Ansari, S., Chandel, A. and Tariq, M. 2021. A comprehensive review on power converters control and control strategies of ac/dc microgrid. IEEE Access, 9: 17998–18015. URL https://api.semanticscholar.org/CorpusID:226694527.

Augustine, S., Mishra, M. K. and Lakshminarasamma, N. 2015. Adaptive droop control strategy for load sharing and circulating current minimization in low-voltage standalone dc microgrid. IEEE Transactions on Sustainable Energy, 6(1): 132–141.

Ballal, M. S., Verma, S. R., Suryawanshi, H. M., Deshmukh, R. R., Wakode, S. A. and Mishra, M. K. 2022. An improved voltage regulation and effective power management by coordinated control scheme in multibus dc microgrid. IEEE Access, 10: 72301–72311.

Bidram, A. and Davoudi, A. 2012. Hierarchical structure of microgrids control system. IEEE Transactions on Smart Grid, 3(4): 1963–1976.

Choi, H. -J., Heo, K. -W. and Jung, J. -H. 2022. A hybrid switching modulation of isolated bidirectional dc-dc converter for energy storage system in dc microgrid. IEEE Access, 10: 6555–6568.

Cunningham, J. J. and Paserba, J. 2022. Distributed generation in 1900? How the edison co. of new york met short-term needs [history]. IEEE Power and Energy Magazine, 20(6): 82–89.

Dhaou, I. S. B., Kondoro, A., Kakakhel, S. R. U., Westerlund, T. and Tenhunen, H. 2022. Internet of Things Technologies for Smart Grid, Volume 2.

Dragičević, T., Lu, X., Vasquez, J. C. and Guerrero, J. M. 2016. DC microgrids—Part II: A review of power architectures, applications, and standardization issues. IEEE Transactions on Power Electronics, 31(5): 3528–3549.

Dragičević, T., Wheeler, P. and Blaabjerg, F. 2018. DC distribution systems and microgrids. The Institution of Engineering and Technology.

Einhorn, M., Conte, F. V., Kral, C. and Fleig, J. 2013. Comparison, selection, and parameterization of electrical battery models for automotive applications. IEEE Transactions on Power Electronics, 28(3): 1429–1437.

Forouzesh, M., Siwakoti, Y. P., Gorji, F. S., Blaabjerg, A. and Lehman, B. 2017. Step-up dc-dc converters: A comprehensive review of voltage-boosting techniques, topologies, and applications. IEEE Transactions on Power Electronics, 32(12): 9143–9178.

Gao, F., Kang, R., Cao, J. and Yang, T. 2018. Primary and secondary control in dc microgrids: A review. Journal of Modern Power Systems and Clean Energy, 7(2): 227–242.

Ghasempour, A. 2019. Internet of things in smart grid: Architecture, applications, services, key technologies, and challenges. Inventions, 4(1). ISSN 2411-5134. URL https://www.mdpi.com/2411-5134/4/1/22.

Guerrero, J. M., Vásquez, J. C. and Teodorescu, R. 2009. Hierarchical control of droop-controlled dc and ac microgrids—A general approach towards standardization. In 2009 35th Annual Conference of IEEE Industrial Electronics, pp. 4305–4310.

Hamidieh, M. and Ghassemi, M. 2022. Microgrids and resilience: A review. IEEE Access, 10: 106059–106080.

He, H., Xiong, R. and Fan, J. 2011. Evaluation of lithium-ion battery equivalent circuit models for state of charge estimation by an experimental approach. Energies, 4(4): 582–598. ISSN 1996-1073. URL https://www.mdpi.com/1996-1073/4/4/582.

Hu, C., Youn, B. D. and Chung, J. 2012. A multiscale framework with extended kalman filter for lithium-ion battery soc and capacity estimation. Applied Energy, 92: 694–704. ISSN 0306-2619. URL https://www.sciencedirect.com/science/article/pii/S0306261911004971.

Kakigano, H., Miura, Y. and Ise, T. 2010. Low-voltage bipolar-type dc microgrid for super high quality distribution. IEEE Transactions on Power Electronics, 25(12): 3066–3075.

Kasper, M., Bortis, D. and Kolar, J. W. 2014. Classification and comparative evaluation of pv panel-integrated dc–dc converter concepts. IEEE Transactions on Power Electronics, 29(5): 2511–2526.

Kazimierczuk, M. K. 2016. Pulse-width modulated DC-DC power converters. Wiley.

Kim, S. and Chou, P. H. 2015. Energy Harvesting with Supercapacitor-Based Energy Storage, pp. 215–241. Springer International Publishing, Cham, 2015. ISBN 978-3-319-14711-6.

Lee, M., Choi, W., Kim, H. and Cho, B. -H. 2015. Operation schemes of interconnected dc microgrids through an isolated bi-directional dc-dc converter. In 2015 IEEE Applied Power Electronics Conference and Exposition (APEC), pp. 2940–2945.

Li, X. and Choe, S. -Y. 2013. State-of-charge (SoC) estimation based on a reduced order electrochemical thermal model and extended kalman filter. In 2013 American Control Conference, pp. 1100–1105.

Lotfi, H. and Khodaei, A. 2017. Ac versus dc microgrid planning. IEEE Transactions on Smart Grid, 8(1): 296–304.

Lu, X., Guerrero, J. M., Sun, K. and Vasquez, J. C. 2014. An improved droop control method for dc microgrids based on low bandwidth communication with dc bus voltage restoration and enhanced current sharing accuracy. IEEE Transactions on Power Electronics, 29(4): 1800–1812.

Misyris, G. S., Doukas, D. I., Papadopoulos, T. A., Labridis, D. P. and Agelidis, V. G. 2019. State-of-charge estimation for Li-ion batteries: A more accurate hybrid approach. IEEE Transactions on Energy Conversion, 34(1): 109–119.

Moayedi, S. and Davoudi, A. 2016. Distributed tertiary control of dc microgrid clusters. IEEE Transactions on Power Electronics, 31(2): 1717–1733.

Mudaliyar, S., Duggal, B. and Mishra, S. 2020. Distributed tie-line power flow control of autonomous dc microgrid clusters. IEEE Transactions on Power Electronics, 35(10): 11250–11266.

O' Keeffe, D., Riverso, S., Albiol-Tendillo, L. and Lightbodyt, G. 2017. Distributed hierarchical droop control of boost converters in dc microgrids. In 2017 28th Irish Signals and Systems Conference (ISSC), pp. 1–6.

Páez, J. D., Frey, D., Maneiro, J., Bacha, S. and Dworakowski, P. 2019. Overview of dc-dc converters dedicated to hvdc grids. IEEE Transactions on Power Delivery, 34(1): 119–128.

Park, J. -D., Candelaria, J., Ma, L. and Dunn, K. 2013. Dc ring-bus microgrid fault protection and identification of fault location. IEEE Transactions on Power Delivery, 28(4): 2574–2584.

Pinomaa, A., Ahola, J. and Kosonen, A. 2011. Power-line communication-based network architecture for LVDC distribution system. In 2011 IEEE International Symposium on Power Line Communications and its Applications, pp. 358–363.

Rangarajan, S. S., Raman, R., Singh, A., Shiva, R C., Kumar, K., Sadhu, P. K., Collins, E. R. and Senjyu, T. 2023. Dc microgrids: A propitious smart grid paradigm for smart cities. Smart Cities, 6(4): 1690–1718. ISSN 2624-6511. URL https://www.mdpi.com/2624-6511/6/4/79.

Sabry, A. H., Shallal, A. H., Hameed, H. S. and Ker, P. J. 2020. Compatibility of household appliances with dc microgrid for pv systems. Heliyon, 6(12): e05699. ISSN 2405-8440. URL https://www.sciencedirect.com/science/article/pii/S2405844020325421.

Salem, M., Jusoh, A., Idris, H. N. R., Das, N. and Alhamrouni, I. 2018. Resonant power converters with respect to passive storage (lc) elements and control techniques—An overview. Renewable and Sustainable Energy Reviews, 91: 504–520. ISSN 1364-0321. URL https://www.sciencedirect.com/science/article/pii/S1364032118302302.

Shaban, M., Ben Dhaou, I., Alsharekh, M. F. and Abdel-Akher, M. 2021. Design of a partially grid-connected photovoltaic microgrid using iot technology. Applied Sciences, 11(24). ISSN 2076-3417. URL https://www.mdpi.com/2076-3417/11/24/11651.

Tradacete, M., Santos, C., Jiménez, J. A., Rodríguez, F. J., Martín, P., Santiso, E. and Gayo, M. 2021. Turning base transceiver stations into scalable and controllable dc microgrids based on a smart sensing strategy. Sensors, 21(4). ISSN 1424-8220. URL https://www.mdpi.com/1424-8220/21/4/1202.

Van Heddeghem, W., Lambert, S., Lannoo, B., Colle, D., Pickavet, M. and Demeester, P. 2014. Trends in worldwide ICT electricity consumption from 2007 to 2012. Computer Communications, 50: 64–76. ISSN 0140-3664.

Weng, C., Sun, J. and Peng, H. 2014. A unified open-circuit-voltage model of lithium-ion batteries for state-of-charge estimation and state-of-health monitoring. Journal of Power Sources, 258: 228–237. ISSN 0378-7753. URL https://www.sciencedirect.com/science/article/pii/S0378775314002092.

Wu, Y., Wu, Y., Cimen, H., Vasquez, J. C. and Guerrero, J. M. 2022. Towards collective energy community: Potential roles of microgrid and blockchain to go beyond p2p energy trading. Applied Energy, 314: 119003.

Xiao, J., Wang, P., Setyawan, L. and Xu, Q. 2016. Multi-level energy management system for real-time scheduling of dc microgrids with multiple slack terminals. IEEE Transactions on Energy Conversion, 31(1): 392–400.

Zhang, L., Wu, T., Xing, Y., Sun, K. and Gurrero, J. M. 2011. Power control of dc microgrid using dc bus signaling. In 2011 Twenty-Sixth Annual IEEE Applied Power Electronics Conference and Exposition (APEC), pp. 1926–1932.

Zhang, Z., Ding, T., Zhou, Q., Sun, Y., Qu, M., Zeng, Z., Ju, Y., Li, L., Wang, K. and Chi, F. 2021. A review of technologies and applications on versatile energy storage systems. Renewable & Sustainable Energy Reviews, 148: 111263–2021. ISSN 1364-0321.

CHAPTER 2

Internet-of-Things-Based Communication in Microgrids

Aron Kondoro [a,*] and *Imed Ben Dhaou* [b,c,d]

1. Introduction

1.1. Overview of microgrids

A microgrid is a small-scale version of the traditional electrical grid that can operate independently or in conjunction with the larger grid (Anvari-Moghaddam et al., 2021; Rahmani-Andebili, 2021; Tavakoli et al., 2018). Unlike the traditional grid's centralized power generation and distribution model, a microgrid is decentralized, making it more flexible and resilient. This makes microgrids particularly useful in providing power to remote or isolated areas or when the main grid is unavailable due to an outage or disaster (Mishra et al., 2020; Peterson et al., 2021).

One of the key benefits o f m icrogrids i s t heir a bility t o p rovide a reliable power source to communities and businesses that may not be connected to the main grid. By generating their electricity locally, microgrids can ensure a steady power supply even in an outage on the main grid. This can be especially important for essential services such as hospitals and emergency services, which need a reliable power source to continue operating.

[a] University of Dar es Salaam, Tanzania.

[b] Department of Computer Science, Hekma School of Engineering, Computing, and Design, Dar Al-Hekma University, Jeddah 22246-4872, Saudi Arabia.

[c] Department of Computing, University of Turku, Finland.

[d] Department of Technology, Higher Institute of Computer Sciences and Mathematics, University of Monastir, Tunisia.

Email: imed.bendhaou@utu.fi.

* Corresponding author: kondoro@udsm.ac.tz

Another advantage of microgrids is their potential to support the integration of renewable energy sources (Akinyele et al., 2018). Because they are decentralized, microgrids can easily incorporate renewable energy technologies such as solar or wind power. This can help to reduce reliance on fossil fuels and improve the sustainability of the power system (Lee et al., 2021).

Additionally, microgrids can provide economic benefits to the communities and businesses they serve. By generating their power locally, microgrids can reduce the need for long-distance transmission of electricity, which can be expensive and inefficient. This can lower consumer electricity costs and provide new economic opportunities for local power generation.

Microgrids are an essential part of the future of the power system. By providing a flexible and resilient power source, microgrids can help ensure a reliable electricity supply to communities and businesses, support the integration of renewable energy, and provide economic benefits. As the demand for clean, reliable, and affordable power grows, microgrids will likely play an increasingly important role in meeting that demand.

1.2. The role of IoT in microgrid communication

The modernization of the utility grid towards grid 4.0, commonly known as the smart grid, mandates the use of information and communication technology (ICT) in the operation, control, and management of the grid. In a conventional grid system, the power is generated in one location, transmitted over a long distance and distributed to customers. This architecture is centralized and uses one-way communication, which endued from blackouts, unreliability, and inefficiency. The pressing need to address climate change has motivated the international communities to substitute fossil oil with renewable energies. The smart grid has been conceived to integrate renewable energies along its distribution line.

Advances in semiconductors, sensors, and wireless communication technologies have enabled the development of a new connectivity system commonly known as the Internet of Things (IoT). In the legacy system, the data is collected by a sensor and treated by a local system. This system has been replaced with a wireless sensor network (WSNs) in which each sensor node is equipped with a sensing unit, a tiny processor for pre-processing, a power unit, and a communication block (Akyildiz et al., 2002). Energy-harvesting techniques can be used to further extend the operation mode of the sensor node.

The block diagram of a wireless sensor node is illustrated in Figure 1, detailing its components and functionalities. Table 1 provides a comprehensive list of the common sensor nodes employed in DC microgrids, offering insights into the diverse range of sensors utilized within this context.

Wireless sensor networks have been used in the utility grid as a low-cost solution for monitoring and diagnostics operations Gungor et al. (2010). The application of the WSNs includes but is not limited to automatic metering reading,

Table 1. Commonly used sensor node in the DC microgrid.

Sensor Node Type	Features	References
Solar Radiation Sensors	Measure sunlight intensity.	(López-Lapeña and Pallas-Areny, 2018)
Voltage and Current Sensors	Monitor voltage and current levels.	(Schlüter et al., 2021)
Temperature Sensors	Track temperature of equipment or building.	(Felipe et al., 2021)
Air Quality Sensors	Measure environmental conditions for optimized energy management.	(Kuncoro et al., 2022)

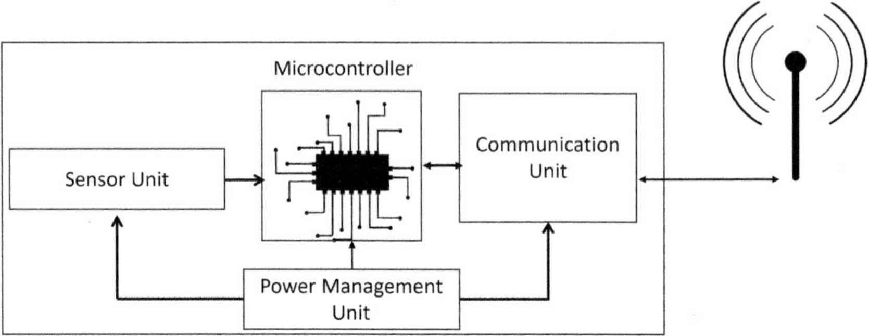

Figure 1. Block diagram of a wireless sensor node.

Table 2. IoT layered architecture.

Layers	Publication
Three layers Perception, network, and application	(Mahmoud et al., 2015)
Four layers, International Telecommunication Union, ITU, reference model Perception, transport, processing, and application	(Kafle et al., 2016)
Five layers Perception, transport, processing, application, and business	(Khan et al., 2012)

remote monitoring, optimization, and equipment fault diagnostics in the transmission and distribution system (Devidas and Ramesh, 2010), power quality monitoring, wide-area situation awareness, and distribution automation (Ogbodo et al., 2017).

The surge of the Internet of Things (IoT) has paved the way for more opportunities to improve the electrical grid. IoT is widely described as a network of physical objects that are embedded with sensors, software, and other technologies for the purpose of connecting and exchanging data with other devices and systems over the internet. Numerous layered architectures for the IoT have been elaborated, which are summarized in Table 2.

The ITU model reference described in ITU-T Y.2060 is pictorially shown in Figure 2, which is composed of four layers: Device or perception layer, network layer, service and application support layer, and application layer. Additionally, the reference model has common management and security capabilities.

By connecting devices such as sensors and smart meters to the internet, the smart grid can better monitor and control the flow of electricity. This can reduce outages, improve energy management, and integrate renewable energy sources. It has also made cloud computing services available for managing smart metre data (Lohrmann and Kao, 2011), grid monitoring and control, and demand-response programs Markovic et al. (2013).

One of the key benefits of the IoT in the smart grid is its ability to provide real-time data on the state of the power system (Dhaou, 2023). This can be used

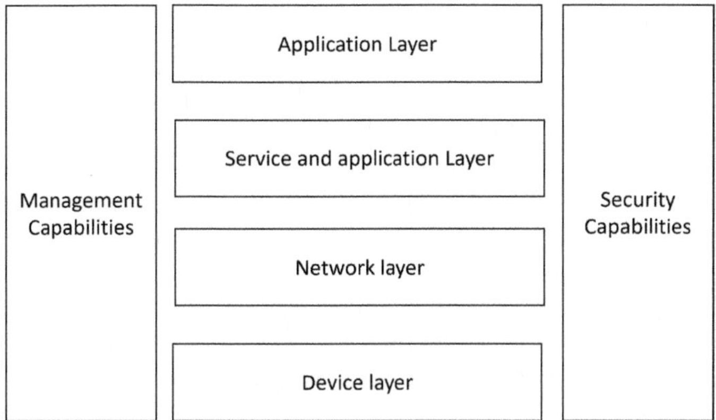

Figure 2. Reference model for the Internet of Things (IoT) eleborated by ITU.

to detect and prevent problems before they occur, such as identifying potential equipment problems or detecting unusual patterns in electricity consumption (Zidi et al., 2023). By proactively addressing potential problems, the smart grid can reduce the likelihood of outages and improve the reliability of the power system.

Another benefit of the IoT in the smart grid is its ability to enable the integration of renewable energy sources. With sensors and other connected devices, the smart grid can more effectively manage the variable output of renewable energy sources such as wind and solar power. This can help to reduce the need for traditional fossil fuel-based power generation, improving the sustainability of the power system.

IoT in the smart grid can also improve customer service and reduce costs for utilities and consumers. Smart grid technologies can help consumers make more informed decisions about their energy consumption by providing real-time electricity usage and pricing information. This can help reduce energy waste and lower electricity bills for consumers while enabling utilities to manage demand more effectively and reduce the need for expensive peak power generation.

IoT in the smart grid can significantly improve the power system's efficiency, reliability, and sustainability. By enabling real-time monitoring and control of the flow of electricity, the smart grid can reduce outages, improve energy management, and enable the integration of renewable energy sources. This can benefit utilities and consumers, making the power system more effective and affordable.

2. IoT communication in microgrids

2.1. Types of IoT communication in microgrids

In microgrids, communication is crucial in enabling efficient and reliable operation. With the advent of the Internet of Things (IoT), various communication technologies have emerged to facilitate data exchange between devices and systems within microgrids (Marzal et al., 2018).

One type of IoT communication commonly used in microgrids is wireless communication. Wireless technologies like Wi-Fi, Bluetooth, and Zigbee provide flexible and convenient device connectivity. These technologies allow devices within the microgrid to communicate wirelessly, enabling real-time data exchange and control. Wireless communication is beneficial when wired connections are impractical or costly.

Another type of IoT communication in microgrids is powerline communication (PLC). PLC utilizes the existing power distribution infrastructure to transmit data signals. By modulating data onto the power lines, devices within the microgrid can communicate. PLC offers the advantage of utilizing the existing power infrastructure, eliminating the need for additional communication cables.

Furthermore, cellular communication is also employed in microgrids. Cellular networks provide wide-area coverage and reliable connectivity, making them suitable for remote monitoring and control of microgrid devices. Cellular communication allows devices within the microgrid to connect to the internet and exchange data with central control systems or cloud platforms.

In addition to these types of IoT communication, microgrids may utilize communication protocols specifically designed for IoT applications. For example, CoAP (Constrained Application Protocol), MQTT (Message Queue Telemetry Transport), and XMPP (Extensible Messaging and Presence Protocol) are commonly used protocols in microgrids. These protocols enable efficient and secure communication between devices and systems, supporting various IoT applications within the microgrid.

CoAP (Constrained Application Protocol) is a lightweight communication protocol for resource-constrained devices and networks. It is commonly used in IoT applications and microgrids for communication between devices, such as sensors and actuators. CoAP uses a simple request-response communication model, which can be used over UDP and TCP.

MQTT (Message Queue Telemetry Transport) is a publish-subscribe communication protocol for low-bandwidth, high-latency networks. It is commonly used in IoT applications and microgrids for communication between devices and systems. MQTT uses a broker-client architecture to manage client communication and ensure that messages are delivered to the correct clients (Arbab-Zavar et al., 2021).

XMPP (Extensible Messaging and Presence Protocol) is a communication protocol for instant messaging and presence information. It is based on XML and is designed to be extensible and flexible. XMPP can be used in microgrids for communication between devices and systems, such as sending commands and receiving status updates.

All of these protocols can be used in microgrids to enable communication between devices and systems, and the choice of which protocol to use will depend on the specific requirements and constraints of the microgrid, such as the data transfer rate, the power requirements of the devices, and the type of communication required.

2.2. IoT platforms, standards, software, and hardware in microgrids

Microgrid communication relies on IoT platforms, standards, software, and hardware to ensure efficient and reliable operation. These technologies enable communication, control, and monitoring within microgrids, contributing to their overall performance and resilience. Several communication standards have been developed for microgrids to ensure compatibility and interoperability between different devices and systems.

One of the key communication standards used in smart grids is the IEEE 802.15.4 standard. This standard defines the communication protocol's physical and media access control (MAC) layer, providing a low-power, low-data-rate wireless communication system. This standard is widely used in smart grids, particularly for applications such as building automation and home area networks.

Another important communication standard used in smart grids is the IEC 61850 standard. This standard defines a common communication protocol for exchanging data between devices and systems in the electrical grid (Albarakati et al., 2022). This allows different smart grid components, such as sensors, meters, and control systems, to communicate with each other and share information.

The Zigbee Alliance has also developed communication standards for smart grids and IoT applications. The Zigbee standards define a low-power, low-cost wireless communication system that is well-suited to the requirements of smart grids. The Zigbee standards are widely used in smart grid applications such as home area networks and building automation.

IoT platforms provide a framework for connecting and managing devices within the microgrid. One commonly used platform is Thingspeak, which is often employed for simulation purposes in power system analysis using Matlab/Simulink (Albarakati et al., 2022). These platforms facilitate data collection, analysis, and control, enabling real-time monitoring and optimising the microgrid's performance.

Software-defined networking (SDN) is another important technology used in microgrid communication. SDN allows for centralized control and management of the network, enabling dynamic and flexible communication within the microgrid (Danzi et al.). It provides the ability to allocate network resources based on real-time conditions and optimize communication performance. Additionally, SDN-based microgrid control architectures have been proposed to enhance resilience against denial-of-service attacks and enable agile reconfiguration of the communication system (Danzi et al.).

Regarding hardware, various devices and components are utilized in microgrid communication. Communication gateways interface the microgrid and external networks, facilitating data exchange and control commands. Sensors and actuators provide real-time data on the microgrid's performance and enable remote control and monitoring (Albarakati et al., 2022). Control systems, such as programmable logic controllers (PLCs), are responsible for managing and coordinating the operation of the microgrid.

The availability of secure, efficient, and reliable communication systems is crucial for microgrids' successful deployment and operation (Kondoro et al., 2021). Building a secure, efficient, reliable communication system remains a challenge. Still, advancements in communication architectures and technologies, such as quantum networks, are being explored to enhance the resilience of microgrids (Tang et al.).

2.3. Implementation of IoT protocols and standards

2.3.1. Steps involved in planning and implementing IoT communication in a microgrid

The planning and implementation of IoT communication in a microgrid involve several steps. Firstly, a thorough assessment of the microgrid's communication requirements and objectives is necessary. This includes identifying the specific data that must be collected and transmitted and the desired communication range, speed, and security (Mina-Casaran et al., 2021).

Once the communication requirements are defined, the next step is to select the appropriate communication technology. This involves considering factors such as the range of communication needed, the speed of data transmission required, and the level of security needed to protect the microgrid's operations (Mina-Casaran et al., 2021). The choice of communication technology will depend on the specific needs and constraints of the microgrid.

After selecting the communication technology, the next step is to design the communication infrastructure. This includes determining the placement and configuration of communication devices within the microgrid, such as sensors, actuators, and gateways (Mina-Casaran et al., 2021). The communication infrastructure should be designed to ensure reliable and efficient data transmission throughout the microgrid.

Once the communication infrastructure is designed, the next step is to deploy and integrate the communication devices into the microgrid. This involves installing and configuring the communication devices and establishing network connections (Mina-Casaran et al., 2021). Ensuring the communication devices are properly integrated with the microgrid's control and monitoring systems is important.

Finally, ongoing monitoring and maintenance of the IoT communication system is essential to ensure its continued performance and reliability. Regular monitoring and troubleshooting can help identify and resolve any issues that may arise, such as communication disruptions or security vulnerabilities (Kondoro et al., 2021).

2.3.2. *Considerations for choosing the right communication technology, including range, speed, and security*

Several considerations must be considered when choosing the right communication technology for a microgrid. One important consideration is the range of communication required. The communication technology should provide sufficient coverage to reach all the devices and components within the microgrid, including those located in remote or hard-to-reach areas (Ben Dhaou et al., 2017).

Another consideration is the speed of data transmission. The communication technology should transmit data at a speed that meets the real-time requirements of the microgrid's control and monitoring systems. This is particularly important for applications that require fast response times, such as fault detection and isolation.

Security is also a critical consideration when choosing a communication technology for a microgrid (Ben Dhaou et al., 2017). The communication technology should provide robust security measures to protect the microgrid's operations from unauthorized access, data breaches, and cyber-attacks. This includes encryption of data transmission, authentication of devices, and secure protocols for communication (Kondoro et al., 2021).

Furthermore, the scalability and flexibility of the communication technology should be considered. The technology should accommodate the future growth and expansion of the microgrid, allowing for the addition of new devices and components without significant modifications to the communication infrastructure.

Overall, the choice of communication technology for a microgrid should be based on a careful evaluation of the range, speed, security, scalability, and flexibility requirements of the microgrid, ensuring that the selected technology can effectively meet the communication needs of the system (Mina-Casaran et al., 2021).

3. Challenges and limitations of IoT communication in microgrids

3.1. *Security concerns*

There are several security concerns to consider when using Internet of Things (IoT) devices in microgrids (Brinckman et al., 2019):

- Unsecured device access: Without proper security measures, unauthorized individuals may be able to access and control IoT devices within the microgrid, potentially leading to disruptions or damage.
- Data breaches: Sensitive data, such as power usage data or customer information, may be at risk of being accessed by unauthorized individuals if proper security measures are not in place to protect it.
- Physical security: Physical access to IoT devices within the microgrid could allow for tampering or damage.

- Malware: IoT devices may be vulnerable to malware attacks, which could lead to disruptions or damage to the microgrid.
- Lack of standardization: The lack of standardization in the IoT industry can make it challenging to ensure that devices are secure and interoperable within a microgrid.

To address these security concerns, it is essential to implement robust security measures, such as secure authentication protocols, encryption, and regular software updates, and to carefully evaluate the security of any IoT devices before incorporating them into a microgrid (Kondoro et al., 2021).

3.2. Interoperability issues

Interoperability refers to the ability of different devices, systems, or applications to work together and communicate effectively. In the context of using Internet of Things (IoT) devices in microgrids, interoperability issues can arise when different devices cannot communicate or exchange data. This can lead to several problems, including

- Incompatibility: IoT devices may use different communication protocols or standards, making working together difficult.
- Limited functionality: If IoT devices cannot communicate with each other, their capabilities may be limited, leading to reduced efficiency and effectiveness within the microgrid.
- Increased complexity: If different IoT devices cannot communicate with each other, it can be more difficult to manage and maintain the microgrid, as additional effort may be required to ensure that different systems are working together effectively.

To address these interoperability issues, it is important to ensure that all IoT devices within a microgrid are compatible and able to communicate. This may involve using standard communication protocols or implementing gateways or other intermediary systems to facilitate device communication.

3.3. Limited range and coverage

In an Internet of Things (IoT) communication system for microgrids, limited range and coverage can be a problem because they can affect the reliability and effectiveness of the communication system.

Microgrids are local energy systems designed to operate independently or in conjunction with the main power grid. They often consist of distributed energy resources such as solar panels, wind turbines, energy storage systems, smart meters, and other IoT devices that monitor and control the microgrid.

IoT communication connects these devices and enables real-time data exchange and control. However, the range and coverage of the communication system can be limited due to various factors, such as the physical distance between devices, the presence of obstacles, and the type of communication technology being used.

Limited range and coverage can impact the microgrid's performance by reducing the amount of data transmitted and received, leading to delays or errors in the communication process. It can also make it more difficult to remotely monitor and control the microgrid, which can impact the overall efficiency and reliability of the system.

To address these issues, microgrid designers can consider using communication technologies with a longer range or higher coverage or deploy additional communication devices to improve the system's coverage. They may also need to consider the layout and placement of the devices to ensure that they are optimally positioned to maximize the range and coverage of the communication system.

3.4. Limited communication bandwidth

In an Internet of Things (IoT) communication system for microgrids, limited communication bandwidth can be a problem because it can affect the reliability and effectiveness of the communication system.

Microgrids are local energy systems designed to operate independently or in conjunction with the main power grid. They often consist of distributed energy resources such as solar panels, wind turbines, energy storage systems, smart meters, and other IoT devices that monitor and control the microgrid.

IoT communication connects these devices and enables real-time data exchange and control. However, the available bandwidth limits the data transmitted and received over the communication system. If the communication system cannot handle the volume of transmitted data, it can result in delays or errors in the communication process.

Limited communication bandwidth can impact the microgrid's performance by reducing the amount of data transmitted and received, leading to delays or errors in the communication process. It can also make it more difficult to remotely monitor and control the microgrid, which can impact the overall efficiency and reliability of the system.

To address these issues, microgrid designers can consider using communication technologies with higher bandwidth or more efficiently using the available bandwidth. They may also need to optimize the data transmission and communication protocols to ensure the communication system is used effectively and efficiently.

4. Strategies to mitigate IoT communication barriers in microgrids

4.1. Encryption and authentication techniques

Encryption and authentication techniques are crucial in ensuring the security and integrity of IoT communication within microgrids. By implementing these techniques, the confidentiality, authenticity, and integrity of data transmitted between IoT devices and the microgrid's communication infrastructure can be safeguarded (Ertürk et al., 2019).

Encryption involves using cryptographic algorithms to convert data into an unreadable format, known as ciphertext, which can only be decrypted by authorized recipients with the corresponding decryption key. This ensures that even if unauthorized individuals access the transmitted data, they cannot decipher its contents. Advanced encryption algorithms, such as Advanced Encryption Standard (AES) or Elliptic Curve Cryptography (ECC), are commonly used to secure IoT communication.

Authentication techniques are employed to verify the identity of IoT devices and ensure that only authorized devices can access the microgrid's communication infrastructure. This helps prevent unauthorized devices from infiltrating the network and compromising its security. Various authentication methods can be used, including digital certificates, public-key infrastructure (PKI), and biometric authentication. These techniques establish trust between devices and the microgrid, ensuring that only legitimate devices can participate in the communication network.

A combination of encryption and authentication techniques can be employed to enhance the security of IoT communication. For example, data can be encrypted using a symmetric encryption algorithm, and the encryption key can be securely exchanged using an asymmetric encryption algorithm during the authentication process. This ensures that data remains confidential and that only authorized devices can access and decrypt the data.

It is important to note that encryption and authentication techniques should be implemented at multiple levels within the microgrid's communication infrastructure. This includes securing communication channels, data storage and transmission, and the interfaces between IoT devices and the microgrid's control and monitoring systems. Regular updates and patches to address vulnerabilities in encryption and authentication protocols should also be applied to ensure the ongoing security of the IoT communication system.

4.2. Standardisation and regulation

The lack of standardization in the IoT industry poses challenges for ensuring the security and interoperability of devices within microgrids. To overcome this challenge, efforts should be made to establish industry-wide standards and regulations for IoT devices. Standardization can ensure that devices meet minimum security requirements and can seamlessly integrate into microgrid communication systems. Regulatory frameworks can also enforce compliance with security standards and provide guidelines for secure IoT deployment (Behjati et al., 2021).

Standardization efforts aim to establish common protocols, interfaces, and frameworks that enable seamless integration and interoperability among IoT devices and systems. These standards ensure that devices from different manufacturers can communicate effectively and securely within the microgrid environment. Standardization also facilitates the development of secure communication protocols, encryption algorithms, and authentication mechanisms that can be universally implemented across IoT devices in microgrids (Al-Fuqaha et al., 2015).

Regulatory frameworks are essential for ensuring the security and privacy of IoT communication within microgrids. Regulations can establish guidelines and requirements for designing, deploying, and operating IoT devices, ensuring they adhere to specific security standards and best practices. Regulatory bodies can also enforce compliance with these standards, conduct audits, and impose penalties for non-compliance, thereby incentivizing the adoption of secure IoT communication practices (Abomhara and Koien, 2014).

Establishing industry-wide standards and regulations promotes trust and confidence in IoT communication within microgrids. It provides a framework for manufacturers, developers, and operators to follow, ensuring that security measures are implemented consistently and effectively. Standardization and regulation also foster innovation by creating a level playing field and enabling interoperability among IoT devices and systems (Centenaro et al., 2021).

Collaboration among industry stakeholders, standardization organizations, and regulatory bodies is crucial for developing and implementing effective standards and regulations. This collaboration ensures that the standards and regulations address IoT communication's needs and challenges in microgrids, considering security, privacy, scalability, and compatibility with existing infrastructure (Silva et al., 2019).

4.3. *Increased range and coverage*

Several solutions and technologies can address the challenges of range and coverage in IoT communication within microgrids. One effective approach is using repeaters or signal amplifiers, which extend the communication range and improve coverage within the microgrid. These devices receive and amplify signals, enabling IoT devices in remote or hard-to-reach areas to connect to the microgrid's communication infrastructure (Marzal et al., 2018).

In addition to repeaters, other technologies can also be employed to enhance range and coverage. Wired technologies like Ethernet (IEEE 802.3) or bus-based technologies like ModBus and ProfiBus can provide reliable and high-speed communication within the microgrid (Marzal et al., 2018). Power-line communication (PLC) is another technology that utilizes existing power lines for data transmission, enabling communication over a wide area without additional wiring (Marzal et al., 2018).

Furthermore, advancements in wireless communication technologies can contribute to extending the range and coverage of IoT communication in microgrids. The deployment of Low-Power Wide-Area Network (LPWAN) technologies, such as LoRaWAN or NB-IoT, can provide long-range and low-power communication capabilities suitable for microgrid applications (Mishra et al., 2020). These technologies enable IoT devices to communicate over large distances while consuming minimal power, making them ideal for remote or distributed microgrid deployments.

Moreover, developing mesh networking protocols can enhance the coverage and reliability of IoT communication within microgrids. Mesh networks consist of interconnected devices that can relay data to extend the range of communication. This self-healing network architecture ensures that even if one device fails or is out of range, data can still be transmitted through alternative paths, improving the overall robustness and coverage of the communication network (Starke et al., 2019).

To optimize range and coverage, it is essential to consider the specific requirements and constraints of the microgrid. Factors such as the size of the microgrid, the geographical layout, and the presence of obstacles or interference sources should be considered when selecting and deploying range-extending technologies. Additionally, proper network planning and optimization techniques can be employed to ensure efficient and reliable communication coverage throughout the microgrid (LeMaster and Hirakawa, 2014).

4.4. *Development of new communication technologies*

The development of new communication technologies is instrumental in addressing the challenges of IoT communication within microgrids. These advancements aim to improve communication systems' efficiency, reliability, and scalability in microgrid environments.

One development area is using 5G technology for IoT communication in microgrids. 5G networks offer higher data transfer rates, lower latency, and increased capacity compared to previous generations of cellular networks. This enables faster and more reliable communication between IoT devices within the microgrid, supporting real-time monitoring, control, and optimization of energy resources.

Another emerging technology is edge computing, which brings computational capabilities closer to IoT devices and data sources. By processing and analyzing data at the network's edge, edge computing reduces latency and bandwidth requirements, enabling faster response times and more efficient utilization of network resources. This is particularly beneficial for time-sensitive applications in microgrids, such as demand response and grid stability control.

As these new communication technologies continue to evolve, it is crucial to consider their compatibility, interoperability, and scalability within microgrid environments. Standardization efforts and regulatory frameworks are vital in ensuring these technologies' seamless integration and secure deployment in microgrids.

5. Advanced technological integration in microgrid IoT communication

5.1. Blockchain technologies in smart microgrids

Blockchain technologies have recently gained significant attention for their potential applications in various industries, including the energy sector. In the context of smart microgrids, blockchain technology offers several benefits, such as enhanced security, transparency, and efficiency. This technology can revolutionize the way energy transactions and data management are conducted in smart microgrids.

One of the key advantages of blockchain technology in smart microgrids is its ability to provide secure and tamper-proof transactions. Alladi et al. discuss using blockchain for secure energy trading in smart grids (Alladi et al., 2019). By utilizing cryptographic algorithms and distributed consensus mechanisms, blockchain ensures that energy transactions are transparent, verifiable, and resistant to tampering. This enhances the security and trustworthiness of energy transactions in smart microgrids, enabling peer-to-peer energy trading without intermediaries.

Blockchain technology also enables improved transparency and traceability of energy transactions in smart microgrids (Dinesha and Patil, 2023). By recording all transactions in a decentralized and immutable ledger, blockchain gives participants a transparent view of energy generation, consumption, and trading activities. This transparency can help promote accountability and trust among participants, facilitating the integration of renewable energy sources and incentivizing energy conservation.

Furthermore, blockchain technology can enhance the efficiency and reliability of energy management in smart microgrids. Energy transactions can be automated and streamlined using smart contracts and self-executing agreements stored on the blockchain. Smart contracts can enable real-time settlement of energy transactions, eliminate the need for manual reconciliation, and reduce administrative costs. This automation and efficiency can contribute to optimising energy distribution and utilization in smart microgrids.

In addition to energy transactions, blockchain technology can facilitate secure and decentralized data management in smart microgrids. Using blockchain-based data management systems, sensitive energy data, such as consumption patterns, grid conditions, and renewable energy generation, can be securely stored, shared, and analyzed (Vaghasana et al., 2023). This can enable data-driven decision-making, grid optimization, and the development of innovative energy services and applications.

Blockchain technology holds great promise for the advancement of smart microgrids. Its ability to provide secure transactions, transparency, and efficiency can revolutionize the energy sector by enabling peer-to-peer energy trading, promoting renewable energy integration, and enhancing data management. However, further research and development are needed to address scalability, interoperability, and regulatory challenges to realise blockchain's potential in smart microgrids fully.

5.2. Machine learning applications

Machine learning has emerged as a valuable tool in various applications, including microgrids. It offers the potential to optimize and enhance the operation of microgrids by leveraging data-driven approaches and intelligent decision-making algorithms.

One application of machine learning in microgrids is intelligent energy management. Chaouachi et al. propose a multiobjective intelligent energy management system for a microgrid (Chaouachi et al., 2013). The study utilizes a fuzzy logic expert system for battery scheduling, a critical aspect of microgrid operation.

Machine learning is used for stability prediction in microgrids. Alazab et al. developed a multidirectional LSTM model for predicting a smart grid's stability, including microgrids (Alazab et al., 2020). The model utilizes long short-term memory (LSTM) networks, a type of recurrent neural network, to capture temporal dependencies and patterns in the data. Training the model on historical data can predict the microgrid's stability and identify potential issues or anomalies that may arise. This enables proactive measures to be taken to maintain the stability and reliability of the microgrid.

Another application of machine learning in microgrids is the detection of bad data. (Huang et al., 2022) present a data-driven approach that combines online machine learning with statistical analysis for sequential detection of microgrid bad data (Huang et al., 2022). The study utilizes supervised online sequential machine learning to approximate the analytical model of the microgrid. This approach enables the detection of anomalies or erroneous data in real-time, allowing for timely corrective actions (Huang et al., 2022).

Machine learning techniques can also be applied to optimize the control and coordination of distributed energy resources (DERs) in microgrids. By analyzing historical data and real-time information, machine learning algorithms can learn patterns and make predictions to optimize the scheduling and dispatch of DERs. This can lead to improved energy efficiency, cost savings, and better integrating renewable energy sources into the microgrid.

Furthermore, machine learning can be utilized for load forecasting in microgrids. Machine learning algorithms can accurately predict future load demand by analyzing historical load data and considering various factors such as weather conditions, holidays, and special events. This information can be used for proactive load management, resource allocation, and grid stability.

6. Summary

Integrating Internet of Things (IoT) technologies in microgrid communication holds immense importance and offers numerous benefits. The application of IoT in microgrids has the potential to revolutionize energy management and enhance these systems' overall performance and efficiency.

One of the key advantages of IoT-enabled communication in microgrids is the ability to collect real-time data from sensors and devices. This data can be used for accurate state estimation, monitoring, and control of microgrid operations. Microgrids can improve situational awareness and make informed real-time decisions by leveraging IoT communication infrastructure.

Furthermore, IoT-based communication enables decentralized control strategies in microgrids. This allows for more efficient and cooperative management of energy resources, leading to optimized energy generation, distribution, and consumption. Decentralized control also enhances the resilience and reliability of microgrids by enabling self-healing and adaptive capabilities.

There are several future developments and opportunities for integrating IoT into microgrid communication. Firstly, there is a need for standardized IoT platforms that are specifically designed for the unique requirements of microgrids. Current IoT standards may not be optimal for developing Internet of Energy (IoE) platforms, which present more demanding challenges.

Research and development efforts should address the cybersecurity concerns associated with IoT-enabled microgrid communication. As microgrids become more interconnected and reliant on IoT technologies, ensuring the security and integrity of data and communication networks becomes crucial.

Moreover, there is a need for further advancements in the interoperability and integration of IoT devices and systems in microgrids. This will enable seamless communication and collaboration between different components and stakeholders in the energy ecosystem.

References

Abomhara, M. and Koien, G. M. 2014, May. Security and privacy in the internet of things: Current status and open issues. In 2014 International Conference on Privacy and Security in Mobile Systems (PRISMS). IEEE.

Akinyele, D., Belikov, J. and Levron, Y. 2018, February. Challenges of microgrids in remote communities: A STEEP model application. Energies, 11(2): 432.

Akyildiz, I., Su, W., Sankarasubramaniam, Y. and Cayirci, E. 2002. A survey on sensor networks. IEEE Communications Magazine, 40(8): 102–114.

Alazab, M., Khan, S., Krishnan, S. S. R., Pham, Q. -V., Reddy, M. P. K. and Gadekallu, T. R. 2020. A multidirectional LSTM model for predicting the stability of a smart grid. IEEE Access, 8: 85454–85463.

Albarakati, A. J., Boujoudar, Y., Azeroual, M., Eliysaouy, L., Kotb, H., Aljarbouh, A., Khalid Alkahtani, H., Mostafa, S. M., Tassaddiq, A. and Pupkov, A. 2022, December. Microgrid energy management and monitoring systems: A comprehensive review. Front. Energy Res., 10: 1097858.

Al-Fuqaha, A., Guizani, M., Mohammadi, M., Aledhari, M. and Ayyash, M. 2015. Internet of things: A survey on enabling technologies, protocols, and applications. IEEE Commun. Surv. Tutor., 17(4): 2347–2376.

Alladi, T., Chamola, V., Rodrigues, J. J. P. C. and Kozlov, S. A. 2019, November. Blockchain in smart grids: A review on different use cases. Sensors, 19(22).

Anvari-Moghaddam, A., Abdi, H., Mohammadi-Ivatloo, B. and Hatziargyriou, N. 2021. Microgrids: Advances in Operation, Control, and Protection. Springer Nature.

Arbab-Zavar, B., Palacios-Garcia, E. J., Vasquez, J. C. and Guerrero, J. M. 2021, September. Message queuing telemetry transport communication infrastructure for grid-connected AC microgrids management. Energies, 14(18): 5610.

Behjati, M., Mohd Noh, A. B., Alobaidy, H. A. H., Zulkifley, M. A., Nordin, R. and Abdullah, N. F. 2021, July. LoRa communications as an enabler for internet of drones towards large-scale livestock monitoring in rural farms. Sensors, 21(15).

Ben Dhaou, I., Kondoro, A., Kelati, A., Rwegasira, D. S., Naiman, S., Mvungi, N. H. and Tenhunen, H. 2017. Communication and security technologies for smart grid. International Journal of Embedded and Real-Time Communication Systems, 8(2): 40–65.

Brinckman, A., Chard, K., Gaffney, N., Hategan, M., Jones, M. B., Kowalik, K., Kulasekaran, S., Ludäscher, B., Mecum, B. D., Nabrzyski, J., Stodden, V., Taylor, I. J., Turk, M. J. and Turner, K. 2019, May. Computing environments for reproducibility: Capturing the "whole tale". Future Gener. Comput. Syst., 94: 854–867.

Centenaro, M., Costa, C. E., Granelli, F., Sacchi, C. and Vangelista, L. 2021. A survey on technologies, standards and open challenges in satellite IoT. IEEE Commun. Surv. Tutor., 23(3): 1693–1720.

Chaouachi, A., Kamel, R. M., Andoulsi, R. and Nagasaka, K. 2013, April. Multiobjective intelligent energy management for a microgrid. IEEE Trans. Ind. Electron., 60(4): 1688–1699.

Danzi, P., Angjelichinoski, M., Stefanović, Č., Dragičević, T. and Popovski, P. (n.d.). Software-defined microgrid control for resilience against denial-of-service attacks. https://doi.org/10.1109/tsg.2018.2879727 (Accessed: 2023-9-13).

Devidas, A. R. and Ramesh, M. V. 2010. Wireless smart grid design for monitoring and optimizing electric transmission in India. In 2010 Fourth International Conference on Sensor Technologies and Applications, pp. 637–640.

Dhaou, I. B. 2023. Design and implementation of an internet-of-things-enabled smart meter and smart plug for home-energy-management system. Electronics, 12(19). Retrieved from https://www.mdpi.com/2079-9292/12/19/4041.

Dinesha, D. L. and Patil, B. 2023, January. Decentralized token exchanges in blockchain enabled interconnected smart microgrids.

Ertürk, M. A., Aydın, M. A., Büyükakkaşlar, M. T. and Evirgen, H. 2019, October. A survey on LoRaWAN architecture, protocol and technologies. Future Internet, 11(10): 216.

Felipe, T. A., Melo, F. C. and Freitas, L. C. G. 2021. Design and development of an online smart monitoring and diagnosis system for photovoltaic distributed generation. Energies, 14(24). Retrieved from https://www.mdpi.com/1996-1073/14/24/8552.

Gungor, V. C., Lu, B. and Hancke, G. P. 2010. Opportunities and challenges of wireless sensor networks in smart grid. IEEE Transactions on Industrial Electronics, 57(10): 3557–3564.

Huang, H., Liu, F., Ouyang, T. and Zha, X. 2022, May. Sequential detection of microgrid bad data via a data-driven approach combining online machine learning with statistical analysis. Front. Energy Res., 10.

Kafle, V. P., Fukushima, Y. and Harai, H. 2016. Internet of things standardization in itu and prospective networking technologies. IEEE Communications Magazine, 54(9): 43–49.

Khan, R., Khan, S. U., Zaheer, R. and Khan, S. 2012. Future internet: The internet of things architecture, possible applications and key challenges. In 2012 10th International Conference on Frontiers of Information Technology, pp. 257–260.

Kondoro, A., Dhaou, I., Tenhunen, H. and Mvungi, N. 2021, October. A low latency secure communication architecture for microgrid control. Energies, 14(19): 6262.

Kuncoro, C. B. D., Asyikin, M. B. Z. and Amaris, A. 2022. Smart-autonomous wireless volatile organic compounds sensor node for indoor air quality monitoring application. International Journal of Environmental Research and Public Health, 19(4). Retrieved from https://www.mdpi.com/1660-4601/19/4/2439.

Lee, H. -J., Vu, B. H., Zafar, R., Hwang, S. -W. and Chung, I. -Y. 2021, January. Design framework of a stand-alone microgrid considering power system performance and economic efficiency. Energies, 14(2): 457.

LeMaster, D. A. and Hirakawa, K. 2014, April. Improved microgrid arrangement for integrated imaging polarimeters. Opt. Lett., 39(7): 1811–1814.

Lohrmann, B. and Kao, O. 2011. Processing smart meter data streams in the cloud. In 2011 2nd IEEE PES International Conference and Exhibition on Innovative Smart Grid Technologies, pp. 1–8.

López-Lapeña, O. and Pallas-Areny, R. 2018. Solar energy radiation measurement with a low–power solar energy harvester. Computers and Electronics in Agriculture, 151: 150–155. Retrieved from https://www.sciencedirect.com/science/article/pii/S0168169918303132.

Mahmoud, R., Yousuf, T., Aloul, F. and Zualkernan, I. 2015. Internet of Things (IoT) security: Current status, challenges and prospective measures. In 2015 10th International Conference for Internet Technology and Secured Transactions (ICITST), pp. 336–341.

Markovic, D. S., Zivkovic, D., Branovic, I., Popovic, R. and Cvetkovic, D. 2013. Smart power grid and cloud computing. Renewable and Sustainable Energy Reviews, 24: 566–577. Retrieved from https://www.sciencedirect.com/science/article/pii/S136403211300227X.

Marzal, S., Salas, R., González-Medina, R., Garcerá, G. and Figueres, E. 2018, February. Current challenges and future trends in the field of communication architectures for microgrids. Renewable Sustainable Energy Rev., 82: 3610–3622.

Mina-Casaran, J. D., Echeverry, D. F. and Lozano, C. A. 2021, March. Demand response integration in microgrid planning as a strategy for energy transition in power systems. IET Renew. Power Gener., 15(4): 889–902.

Mishra, S., Anderson, K., Miller, B., Boyer, K. and Warren, A. 2020, April. Microgrid resilience: A holistic approach for assessing threats, identifying vulnerabilities, and designing corresponding mitigation strategies. Appl. Energy, 264(114726): 114726.

Ogbodo, E. U., Dorrell, D. and Abu-Mahfouz, A. M. 2017. Cognitive radio based sensor network in smart grid: Architectures, applications and communication technologies. IEEE Access, 5: 19084–19098.

Peterson, C. J., Van Bossuyt, D. L., Giachetti, R. E. and Oriti, G. 2021, September. Analyzing mission impact of military installations microgrid for resilience. Systems, 9(3): 69.

Rahmani-Andebili, M. 2021. Design, control, and operation of microgrids in smart grids. Springer Nature.

Schlüter, M., Gentejohann, M. and Dieckerhoff, S. 2021. High bandwidth on-board dc voltage and current measurements of the main inverter of an electric vehicle. In 2021 IEEE Vehicle Power and Propulsion Conference (VPPC), pp. 1–6.

Silva, J. D. C., Rodrigues, J. J. P. C., Saleem, K., Kozlov, S. A. and Rabelo, R. A. L. 2019. M4DN.IoT-A networks and devices management platform for internet of things. IEEE Access, 7: 53305–53313.

Starke, M., Herron, A., King, D. and Xue, Y. 2019, January. Implementation of a publish-subscribe protocol in microgrid islanding and resynchronization with self-discovery. IEEE Trans. Smart Grid, 10(1): 361–370.

Tang, Z., Zhang, P., Krawec, W. O. and Wang, L. n.d. Quantum networks for resilient power grids: Theory and simulated evaluation. https://doi.org/10.1109/tpwrs.2022.3172374 (Accessed: 2023-9-13).

Tavakoli, M., Shokridehaki, F., Funsho Akorede, M., Marzband, M., Vechiu, I. and Pouresmaeil, E. 2018, September. CVaR-based energy management scheme for optimal resilience and operational cost in commercial building microgrids. Int. J. Electr. Power Energy Syst., 100: 1–9.

Vaghasana, M. 2023. Faculty of Science and Technology, RK University, Rajkot Gujarat, India, & Suliya, D. (2023, March). Review on precise of blockchain technology. International Journal of Scientific Research in Engineering and Management, 07(03).

Zidi, S., Mihoub, A., Mian Qaisar, S., Krichen, M. and Abu Al-Haija, Q. 2023. Theft detection dataset for benchmarking and machine learning based classification in a smart grid environment. Journal of King Saud University - Computer and Information Sciences, 35(1): 13–25. Retrieved from https://www.sciencedirect.com/science/article/pii/S1319157822001562.

CHAPTER 3

Blockchain Technology
for DC Microgrids

Sarra Namane,[a,*] *Imed Ben Dhaou*[b,c,d] and *Ahmim Marwa*[a]

1. Introduction

The increasing adoption of renewable energy sources, such as solar panels and wind turbines, created a need for more efficient ways to integrate these intermittent energy sources into the grid. DC power is often used in renewable energy systems, which makes it a natural choice for renewable energy microgrids. In addition, DC power distribution is inherently more efficient than traditional AC power distribution over short distances, which makes it appealing for localized energy systems such as microgrids. These microgrids are small-scale localized energy distribution networks that offer the promise of enhanced resilience, efficiency, and integration of renewable energy sources. However, their effective operation poses multifaceted challenges, ranging from energy management to trust and security in transactions.

On the contrary, blockchain technology, initially conceived as the foundational technology for cryptocurrencies such as Bitcoin, is a decentralized and distributed ledger system that provides transparency, security, and trust in peer-to-peer transactions. Its potential application in the domain of DC microgrids offers promising solutions to the challenges they face.

[a] Networks and System Laboratory, Badji Mokhtar University, Annaba, Algeria.

[b] Department of Computer Science, Hekma School of Engineering, Computing, and Design, Dar Al-Hekma University, Jeddah 22246-4872, Saudi Arabia.

[c] Department of Computing, University of Turku, Finland.

[d] Department of Technology, Higher Institute of Computer Sciences and Mathematics, University of Monastir, Tunisia.

Email: imed.bendhaou@utu.fi.

* Corresponding author: naamanesara2005@yahoo.fr

DC microgrids generate large volumes of data from various sensors and meters. The blockchain can securely store and manage these data, ensuring data integrity and preventing unauthorized access. Furthermore, verifying the source of energy generation, especially for renewable sources like solar panels, is vital. Blockchain can track the origin of energy generation, providing consumers with the assurance that their energy comes from clean sources.

In addition, blockchain can help integrate diverse energy resources, including distributed energy resources (DERs), such as batteries and electric vehicles, into the DC-microgrid. It can efficiently manage energy supply and demand, reducing grid congestion. Furthermore, blockchain can act as a common platform for interoperability among various components within the DC microgrid, including inverters, meters, and energy storage systems.

However, it is important to note that implementing blockchain in DC microgrids also comes with challenges, such as scalability, energy consumption (in the case of Proof of Work blockchains), and regulatory considerations. Careful planning and consideration of these factors are essential when integrating blockchain technology into dc microgrid systems.

This chapter explores the intersection of blockchain technology and DC microgrids, investigating how this synergy can enable efficient energy management, improve grid reliability, promote renewable energy integration, and foster decentralized governance. Then, the chapter presents the challenges of implementation of Blockchain in DC microgrids and the possible solutions.

1.1. Overview of blockchain technology

Blockchain technology is a revolutionary new way to transfer and store data in a secure and decentralized manner . The foundations of the blockchain can be traced back to the late 20th century, when researchers like Stuart Haber and W. Scott Stornetta explored cryptographic techniques for timestamping digital documents to prevent backdating or tampering.

Blockchain, when defined according to its structural components, can be described as a decentralized digital ledger technology that enables secure and transparent record-keeping of transactions or data in the form of blocks. These blocks are connected through a cryptographic hash function in chronological order, forming a continuous chain.

What makes the blockchain unique is its decentralized nature, using a distributed network system of nodes linked together to verify transactions on the ledger.

This technology relies on cryptographic principles because when a new block is added to the chain, the nodes of the network verify its validity by recalculating its hash. They also check that the previous hash of the new block matches the hash of the last block in the existing chain. This chaining of blocks through their cryptographic hashes ensures the integrity and immutability of the blockchain.

Furthermore, consensus algorithms are used to validate and agree on the content of a new block before it is added to the chain.

Finally, once a block is added to the chain, it becomes extremely difficult to alter its content due to the cryptographic connections and consensus rules. This immutability is crucial to ensure the integrity and trustworthiness of the data recorded on blockchains.

1.2. Importance of blockchain technology in DC-microgrids systems

DC microgrids, like any energy infrastructure, face several challenges that need to be addressed to ensure their reliable and safe operation. First, they are susceptible to cyberattacks, just as any other digital system. Furthermore, malicious actors can attempt to disrupt energy distribution, manipulate energy transactions, or gain unauthorized access to control systems. Furthermore, while DC microgrids are well suited for certain applications like data centers and remote areas, scaling them up to cover larger geographic regions or entire cities can be complex and costly. Finally, ensuring that different components, devices, and systems within the DC microgrid can communicate and trade energy seamlessly can be challenging. Integrating blockchain technology into DC microgrids holds great promise in overcoming the limitations of previous DC microgrids.

The blockchain eliminates the need for a central authority, allowing peer-to-peer energy transactions within the microgrid. This decentralization increases energy resilience and reduces the risk of a single point of failure. Furthermore, the cryptographic features of the blockchain enhance the resilience of the microgrid infrastructure against cyber threats. It makes it difficult for malicious actors to manipulate or disrupt energy transactions.

In addition, blockchain technology can enable real-time monitoring of energy generation, consumption, and distribution, making it easier to balance supply and demand within the microgrid.

Finally, as the microgrid grows or additional participants join, smart contracts can easily scale to accommodate more energy producers and consumers without the need for intermediaries. In addition, smart contracts can be customized to accommodate various energy trading models and help balance the microgrid by incentivizing energy producers to supply energy when needed.

Integrating blockchain in DC microgrids empowers both individual consumers and the broader grid network by paving the way for more resilient, sustainable, and efficient energy systems. This empowers smart contracts to automate energy trading, minimizing human error, and streamlining processes for increased efficiency. Moreover, blockchain safeguards energy transactions through tamper-proof record-keeping, preventing fraudulent activity, and ensuring fair billing. This revolutionary technology effectively addresses many security challenges inherent in distributed energy systems, making DC microgrids more resilient to cyber threats and unauthorized access, ultimately revolutionizing the way we manage and distribute energy.

2. Blockchain technology

2.1. Definition and components of blockchain technology

Blockchain represents a decentralized ledger designed for the efficient storage of digital cash transactions between two parties (Namane and Ben Dhaou, 2022). These transactions are recorded as an expanding series of entries known as blocks. These blocks are resistant to any alterations and offer a permanent and verifiable record. Typically, a collective of users connected through a peer-to-peer (P2P) network is responsible for verifying the ledger entries. To effect changes within these blocks, a consensus must be reached among more than half of the network's users (Mohanta et al., 2019). In blockchain, a block functions as the data structure for the storage of transaction records. As illustrated in Figure 1), it is made up of two essential elements: the head of the block and the body of the block. The header of the block comprises the following fields:

- Block Version: defines requirements for block validation.
- nBits: sets the minimum requirement for a valid block hash.
- Time Stamp: gives the current time in seconds.
- Merkle Tree Root Hash: records of the hash value of each block's transactions.
- Nonce: It is a numerical value that begins at 0 and increases with each hash calculation.
- Parent Block Hash: references the preceding block. The block body includes a transaction counter and the transactions. The block's ability to store transactions is constrained by both the block size and the size of each transaction it holds.

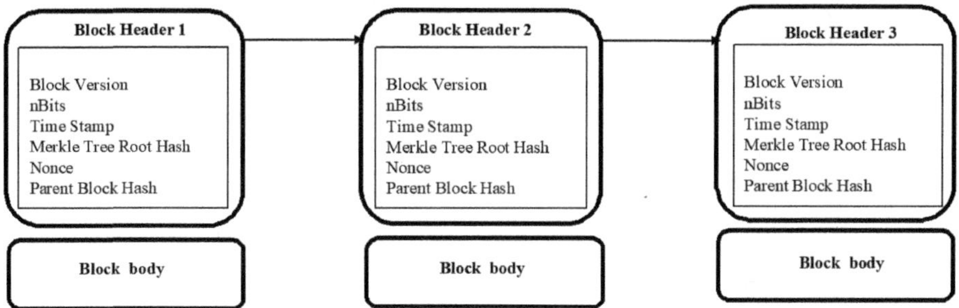

Figure 1. Blockchain structure.

2.2. Types of blockchain technology

Blockchains come in various types to cater to different needs. Permissionless blockchains (Bhutta et al., 2021), often public, such as Bitcoin, allow anyone to participate, read, write, and verify transactions without requiring approval. Permissioned blockchains, on the other hand, restrict access to specific users or organizations, ensuring tighter control over the network. Public blockchains, such as Ethereum, are open to all and transparent, while private blockchains limit access to authorized participants, providing enhanced privacy and security. Hybrid blockchains combine elements of public and private blockchains, offering flexibility in terms of transparency and control, making them a versatile choice for various applications. Figure 2 presents an example of each type.

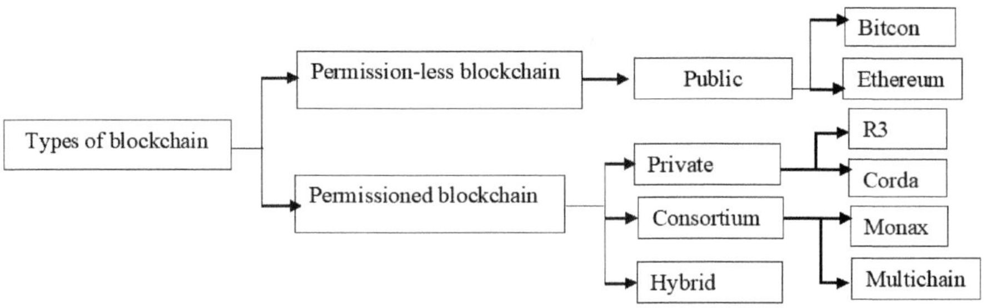

Figure 2. Blockchain types.

2.3. Key features of blockchain technology

Blockchain technology has several key features that make it unique and highly valuable in various industries. First, it provides unparalleled security (Zeng et al., 2020) due to its decentralized and immutable ledger, making it exceptionally resistant to fraud and tampering. Secondly, transparency is a fundamental characteristic since every transaction is meticulously recorded and visible to all participants in the network, thus ensuring trust and accountability. Furthermore, blockchain promotes efficiency by eliminating the need for intermediaries, leading to decreased transaction costs and accelerated processes. It also facilitates trustless transactions, allowing parties to engage in exchanges without relying on a central authority. Lastly, this technology enables the implementation of smart contracts, which are self-executing agreements with predefined rules, further automating processes and diminishing the dependence on intermediaries. Collectively, these characteristics position the blockchain as a transformative tool with great potential in sectors such as finance, supply chain management, healthcare, and beyond (Musa et al., 2023). Figure 3 summarizes the key application of blockchain technology.

3. Use cases of blockchain technology in DC-microgrid systems

This section explores the diverse and innovative use cases of blockchain technology within the realm of DC-microgrids. Blockchain, a decentralized and secure

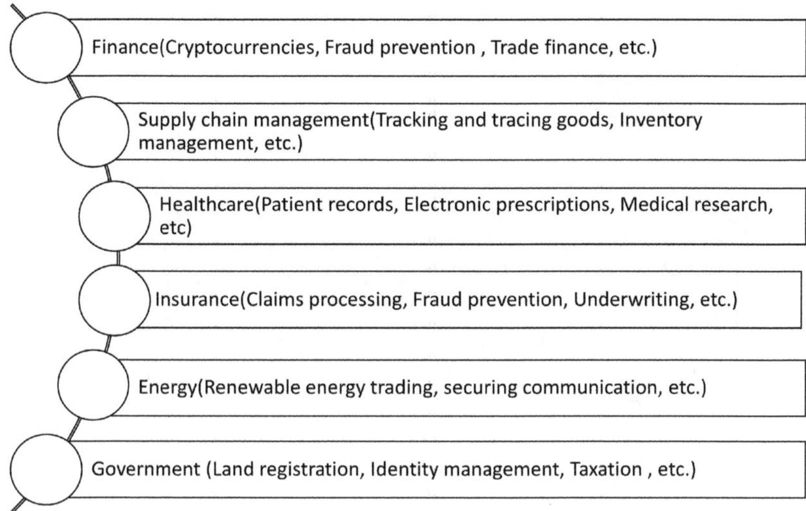

Figure 3. Key application domains of blockhain technology.

ledger system, has the potential to revolutionize the way we manage, monitor, and optimize DC-microgrid systems. From improving energy trading and supply chain management to ensuring data integrity and cybersecurity, the integration of blockchain technology opens up a multitude of possibilities in the field of distributed energy generation and consumption. The following subsections present these applications.

3.1. Energy trading

The application of blockchain technology to energy trading in DC microgrids is a groundbreaking and revolutionary method of overseeing energy distribution and transactions within localized power networks. This emerging technology utilizes the fundamental principles of blockchain, a secure and decentralized digital ledger, to enable transparent, efficient, and secure energy transactions among participants within DC microgrids.

Given a local community where numerous homes are equipped with rooftop solar panels (illustrated in Figure 4). These panels produce surplus electricity during daylight hours when the sun is shining, yet they may not generate sufficient power to meet the community's energy needs after sunset. To improve energy efficiency and decrease dependence on the primary power grid, the community establishes a DC microgrid and leverages blockchain technology for energy exchange.

Each home has a digital wallet associated with its blockchain identity. Furthermore, IoT devices and smart meters are used on each home's electrical system to provide real-time data on energy consumption and production, ensuring

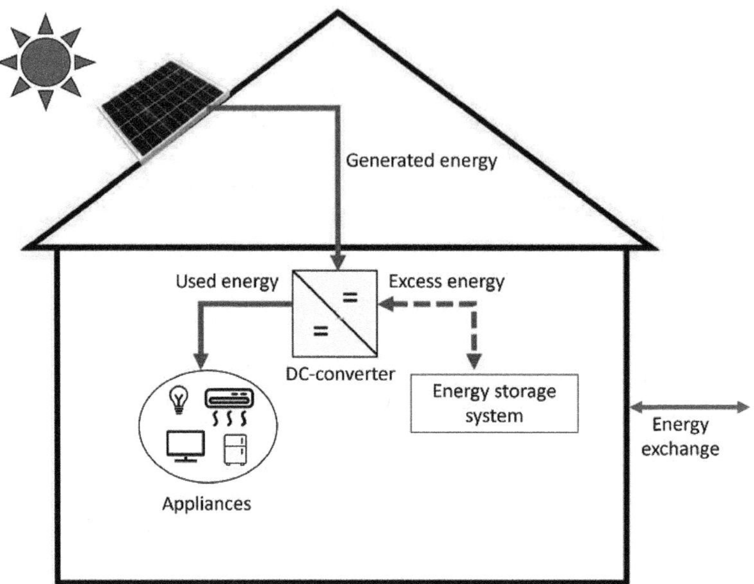

Figure 4. An illustration of energy flow in PV house.

accurate transactions. It is obvious that homes that generate more energy than they consume have surplus energy. This surplus energy is converted into digital tokens on the blockchain. But homes that need more electricity than their solar panels produce can purchase energy tokens from homes with surplus energy. This last process is known as peer-to-peer trading. It is usually achieved using smart contracts (Pee et al., 2019), self-executing agreements based on predefined rules, automating various aspects of energy trading, such as pricing, payment, and verification. The smart contract includes details such as the amount of energy to be transferred, the price per token, and the duration of the energy transfer agreement. Once the smart contract conditions are met (e.g., payment in tokens received), the surplus energy is transferred from the seller's wallet to the buyer's wallet on the blockchain.

In addition, blockchain can trace the origin of energy, making it possible to verify whether electricity comes from renewable sources or not. This can be essential to meet sustainability goals and incentives. In addition, the blockchain automatically calculates the total energy traded and settles the payment in tokens between homes.

Furthermore, the blockchain continuously tracks the energy supply and demand within the microgrid, allowing for efficient balancing of resources and ensuring that everyone's energy needs are met.

In conclusion, blockchain-enabled energy trading in a DC microgrid can enable communities to efficiently share locally generated renewable energy, reduce energy costs, and contribute to a more sustainable and resilient energy system.

3.2. Decentralization

In conventional DC microgrids, the generation, consumption and distribution of electricity are under the control of centralized authorities, which restricts flexibility and may result in inefficiencies. Blockchain technology, on the other hand, establishes a decentralized framework that empowers peer-to-peer energy trading between interconnected devices such as solar panels, batteries, and electric vehicles within the Dc microgrid.

In a decentralized energy trading system for DC microgrids based on blockchain technology, the buying and selling of electricity is controlled by a distributed network of participants rather than a central authority. An example is illustrated in Figure 5.

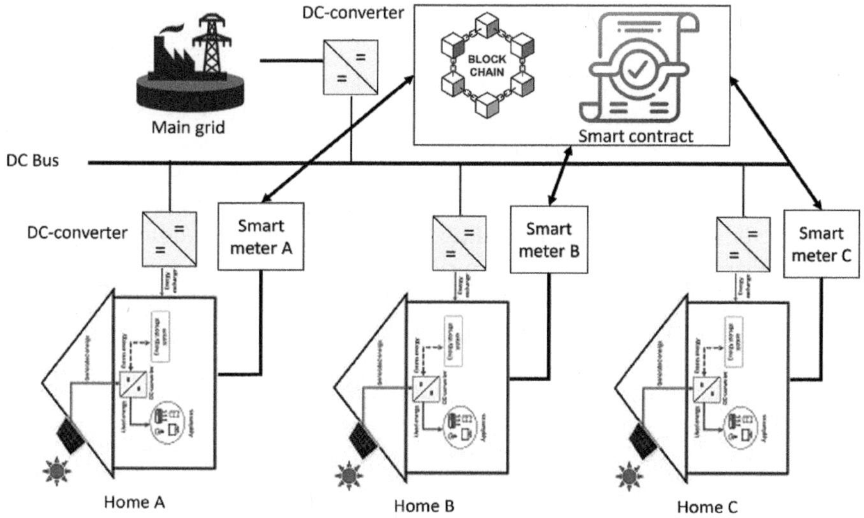

Figure 5. Blockchain enabled energy trading in DC-microgrids.

Let us consider a scenario in which a residential DC-microgrid where several homes have solar panels and energy storage systems (depicted in Figure 4). These homes are part of a blockchain-based energy trading system. Each home is equipped with a smart meter. These meters are capable of measuring electricity consumption and generation at a granular level, up to individual devices, if needed. They also have communication capabilities to send and receive data. In addition, each smart meter continuously collects data on electricity consumption and generation. These data are then time-stamped and securely stored on a blockchain network. In addition, a smart contract is created on the blockchain to define the terms and conditions of energy trading. For example, it could specify the price of electricity, the duration of the contract, and the maximum and minimum amounts of energy that can be traded. For instance, if Home B's smart meter detects a low battery state while Home A's smart meter simultaneously identifies an excess energy generation, the smart meter initiates a transaction specifying the surplus energy amount. Subsequently, Home B's smart meter must trigger a smart contract on the blockchain, outlining the required energy amount, the proposed price, and the transaction's time duration.

The blockchain network then verifies Home B's request and checks the availability of energy from Home A. Upon approval, the blockchain network sends a signal to Home A's smart meter, instructing it to provide the requested amount of DC electricity to Home B's smart meter. Additionally, both smart meters accurately measure the transferred energy and securely record this data in the blockchain ledger.

Once the energy transfer has been completed, the blockchain automatically processes the payment from Home B to Home A based on the agreed-upon price and energy measurement data. This payment is conducted in cryptocurrency or another mutually agreed upon form of payment.

3.3. Security and privacy

Blockchain technology plays a pivotal role in improving the security of DC microgrids. These localized small-scale energy distribution networks are susceptible to various vulnerabilities, including cyberattacks and unauthorized access. Furthermore, numerous factors, such as diverse information resources, sensitive sensors, and extensive interactions both within the DC microgrid (DCMG) and between the DCMG and the main network, along with the need for precise time synchronization and minimized communication delays, present distinct challenges in establishing a reliable and secure operational strategy for DC microgrids (Dehghani et al., 2021).

In blockchain enabled energy trading system, every household's energy consumption and production data is collected, monitored, and transmitted via a smart meter. This meter records both the energy generated by solar panels and the energy consumed by household appliances.

Ensuring the secure transfer of these data is of paramount importance because unauthorized access to this information can lead to significant repercussions.

The previously mentioned concern can be effectively addressed with the application of blockchain technology, as it provides robust safeguards against unauthorized alterations of data related to energy transactions and DC-microgrids operations. Moreover, it allows for granular control over data sharing. Users can grant specific permissions for data access, ensuring that sensitive information is shared only with authorized parties.

Additionally, each smart meter in the DC microgrid is assigned a unique digital identity and possesses a private key. Smart contracts can be programmed to verify the authenticity of these identities against a predefined list of authorized meters, ensuring that only legitimate ones participate in energy transactions within the dc-microgrid (illustrated in Figure 6). In addition, these contracts facilitate secure data exchange by requiring meters to sign transactions with their

```
pragma solidity ^0.8.0;
contract MeterIdentityVerification {
// Mapping to store authorized smart meters identities
   address public owner; // Address of the contract owner
   mapping(address => bool) public authorizedMetersIdentities;
// Constructor to set the contract owner
   constructor() { owner = msg.sender; // The deployer of the contract is the owner }
// Modifier to restrict access to the contract owner
   modifier onlyOwner() {
   require(msg.sender == owner, "Only the contract owner can perform this operation"); }
// Function to add an authorized smart meter identity
   function addAuthorizedmeterIdentity(address _meteridentity) public onlyOwner {
   authorizedmeterIdentities[_meteridentity] = true; }
// Function to remove an authorized smart meter identity
   function removeAuthorizedmeterIdentity(address _meteridentity) public onlyOwner {
   authorizedmeterIdentities[_meteridentity] = false; }
// Function to check if a smart meter identity is authorized
   function isAuthorized(address  meteridentity) public view returns (bool) {
   return authorizedmeterIdentities[_meteridentity];  } }
```

Figure 6. An example of smart contract that checks smart meter identity.

private keys, which are then verified by the contract to confirm the authenticity of the data, preventing data tampering or fraudulent reporting.

4. Future of blockchain technology in DC-microgrid systems

4.1. Potential developments and advancements of blockchain technology

Blockchain technology holds significant importance for a variety of reasons, having a significant impact on numerous industries and aspects of the digital world. However, several challenges must be addressed to successfully implement blockchain technology (Hakak et al., 2021). Furthermore, blockchain has shown significant promise and there are several potential developments and advances that could shape its future.

Blockchain networks often grapple with the limitation of handling a high volume of transactions per second, necessitating scalability solutions for enhanced efficiency. One effective approach involves implementing off-chain solutions like the Lightning Network (Khan and State, 2020) or state channel (Negka and Spathoulas, 2021), which can adeptly manage numerous microtransactions without burdening the primary blockchain. Furthermore, exploring sharding techniques (Hashim et al., 2022), which entail dividing the blockchain network into smaller self-sufficient shards capable of processing their transactions, can significantly boost throughput.

As the blockchain ecosystem continues to expand, the need for different blockchains to communicate and collaborate becomes paramount (Scheid et al., 2019). Interoperability facilitates the seamless transfer of assets, data, and smart contracts between blockchains, fostering a more interconnected and versatile decentralized ecosystem. Existing interoperability solutions focus mainly on linking two permissionless ledgers or as many ledgers as possible, encompassing both permissionless and permissioned ledgers, to establish a network of interconnected ledgers (Koens and Poll, 2019). Nevertheless, it is worth noting that permissioned ledgers might have reservations about disclosing their state within a permissionless ledger. Consequently, there is a pressing need for additional research to explore interoperability between permissionless and permissioned ledgers, with a specific focus on establishing secure connections while safeguarding the privacy attributes of these respective blockchain networks.

Regarding blockchain security, it is important to note that most blockchains operate on public ledgers, which means that all transaction data is visible to anyone. Although transaction details are pseudonymous, they can often be traced, potentially revealing user identities. Additionally, analyzing the transaction history of the blockchain can sometimes allow attackers to link multiple transactions to a single user or entity, thus compromising privacy. Furthermore, smart contracts on blockchains like Ethereum are publicly readable, raising privacy concerns, especially if sensitive data is stored in them. Additionally, these contracts face several issues, including vulnerabilities such as reentry attacks, integer overflows, and unchecked ownership, which can be exploited by malicious actors, resulting in significant financial losses.

To mitigate the security issues mentioned above, enhanced blockchain privacy features like zero-knowledge proofs (Sun et al., 2021) and confidential transactions (Yuen et al., 2020) are being integrated to protect user data while maintaining transparency. In addition, the use of formal verification tools and methods to mathematically verify the correctness of smart contracts is strongly recommended. Moreover, it is essential to implement proper access control mechanisms to ensure that only authorized users or contracts can execute sensitive functions within smart contracts.

4.2. Impact of blockchain technology on DC-microgrid systems

Traditional DC-microgrid energy trading historically relied on centralized systems and intermediaries such as utilities or energy brokers. Transactions were documented in centralized databases and trust was placed in these third-party entities. For example, in DC-microgrid energy trading, a traditional utility company or energy exchange acts as an intermediary between energy producers (such as power plants) and consumers (homes, businesses, etc.). The utility company sets the prices for energy based on various factors, such as supply, demand, and regulatory considerations. Consumers purchase energy from the utility at these established rates.

In contrast, blockchain-based energy trading in DC microgrids uses distributed ledger technology to record and authenticate energy transactions, employing decentralized consensus mechanisms to ensure transparency and security. In that case, the blockchain enables peer-to-peer energy transactions. Imagine a home with solar panels on the roof of the house that generate excess electricity. With blockchain energy trading, the home owner sell this surplus energy directly to his neighbors or nearby businesses.

Furthermore, blockchain introduces a higher level of trust and transparency by recording all transactions on an immutable ledger visible to all participants, preventing any single entity from manipulating data. On the contrary, traditional energy trading often relies on intermediaries for trust and may lack the same degree of transparency, requiring participants to trust centralized entities for accurate transaction recording and settlement.

Furthermore, blockchain-enabled energy trading in DC-microgrids incorporates the use of smart contracts, which are self-executing agreements with predefined rules. These contracts streamline the energy trading process, encompassing pricing, settlement, and delivery, thus reducing the need for manual intervention.

In contrast, traditional energy trading involves manual negotiations for contracts, billing, and settlement processes, which can be time-consuming and prone to errors.

Moreover, blockchain's decentralized nature enhances the resilience of DC-microgrids, as it eliminates a single point of failure. The immutability of data improves security against tampering and fraud. On the contrary, traditional systems may be more vulnerable to centralized attacks or data breaches due to their reliance on centralized databases and trust in single entities.

Integration of blockchain technology into DC microgrids also presents potential social and environmental impacts. Socially, it could improve access to affordable and reliable energy, especially in underserved communities, while empowering individuals to actively participate in energy markets, potentially reducing energy poverty and fostering community involvement in sustainable practices. The transparency offered by the blockchain could strengthen consumer trust, ensure fair and secure energy transactions, and promote social equity. Environmentally, blockchain integration may encourage the use of renewable energy sources, fostering a cleaner energy mix and reducing reliance on fossil fuels. Additionally, it could incentivize energy efficiency practices due to more transparent and efficient energy use within microgrids, contributing to a reduction in the carbon footprint and overall environmental impact. However, realizing these impacts hinges on factors such as the extent of blockchain implementation, the regulatory environment, technology adoption, and community engagement. It is also essential to consider the environmental footprint associated with the manufacturing and energy consumption of blockchain technology to assess its overall environmental impact.

In addition, it has the potential for significant cost-effectiveness by streamlining transactions and reducing intermediary costs in energy trading. This efficiency can reduce overall operating expenses, which benefits both producers and consumers. As for revenue streams, the implementation of blockchain may create new opportunities through innovative market structures, allowing for peer-to-peer energy sales and potential income from transaction fees. Furthermore, enhanced data security and trust facilitated by blockchain could attract investment and foster more efficient grid operations, potentially generating more income.

In general, blockchain energy trading in DC microgrids offers greater decentralization, transparency, automation, and potential cost savings compared to traditional energy trading systems.

4.3. Role of blockchain technology in creating sustainable energy systems

In the realm of DC (direct current) microgrid systems, a sustainable energy system encompasses an intricately designed and interconnected network for generating, distributing, and consuming energy. This system places a paramount emphasis on environmentally and economically viable practices while ensuring unwavering resilience and reliability. Such a system boasts several advantages, including:

- Diminished greenhouse gas emissions, due to the clean and renewable nature of solar energy.
- Augmented energy resilience as communities become less dependent on distant power grids.
- Reduced long-term energy costs, owing to the abundance of free sunlight, reducing dependence on costly fossil fuels.
- Enhanced air quality and reduction of health risks due to the elimination of fossil fuel-based energy sources.

Given their profoundly positive environmental impacts, the creation and efficient utilization of these systems have risen to the forefront as a top priority.

Blockchain technology can help DC-microgrids address the intermittency of renewable energy sources by providing solutions that improve grid stability, energy management, and flexibility. It can facilitate the integration of energy storage solutions, such as batteries, into the DC-microgrid. Smart contracts on the blockchain can manage the charging and discharging of these batteries to store excess energy when renewable sources are abundant and release it when there is a shortage, helping to bridge intermittent gaps.

Furthermore, when renewable energy generation is low, the DC-microgrid can use blockchain to signal energy-intensive devices to temporarily reduce their consumption, helping to balance supply and demand.

The blockchain can also be used to tokenize renewable energy production. Renewable energy producers can earn tokens to generate clean energy, and consumers can use these tokens to pay for their energy consumption. This incentivizes the use of renewables. Furthermore, to ensure that the manufacturing and transportation of components of renewable technologies are as eco-friendly as possible, blockchain can be used to track the supply chain of components and renewable energy equipment.

By integrating blockchain technology into DC-microgrids, these systems can better manage the intermittent nature of renewable energy sources, optimize energy usage, and ensure a reliable and sustainable energy supply even during periods of low renewable energy production. It ensures that renewable sources are prioritized, leading to reduced dependence on fossil fuels, lower greenhouse gas emissions, and a more sustainable energy system.

5. Case studies of blockchain technology in DC-microgrid systems

5.1. Peerp-to-peer electricity trading

The Brooklyn microgrid is an example of a very successful community microgrid that uses a decentralized architecture. It is the first recognized blockchain technology that uses peer-to-peer energy transactions using blockchain technology. As shown in Figure 7, the microgrid consists of consumers, producers, community solar, and energy providers. Each prosumer is connected to the microgrid using a smart meter.

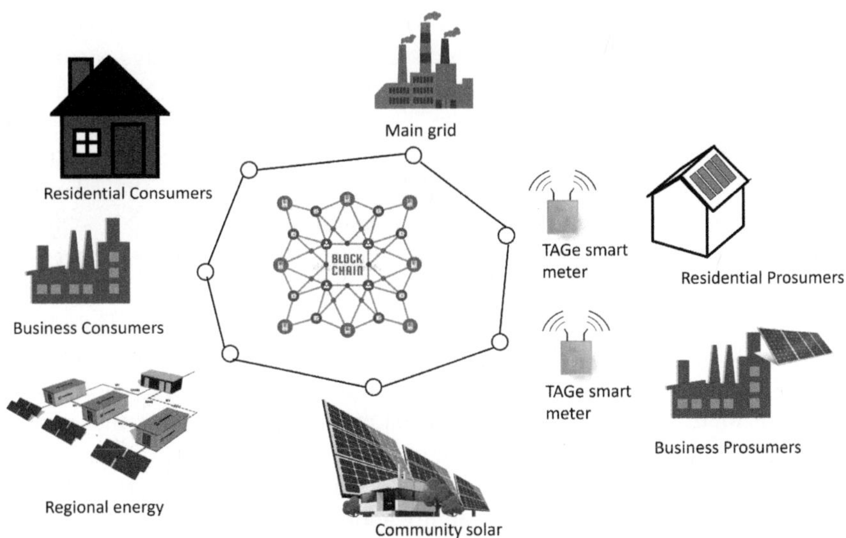

Figure 7. Overview of the Brooklyn microgrid.

5.2. Security

As explained in Chapter 1, hierarchical control is used on DC microgrids. The upper levels, Level 2 and Level 3, rely heavily on a perfect communication link, which can be realized using the IoT communication protocol. The increasing number of attacks on critical infrastructure in recent years made availability, confidentiality, and integrity a viable design consideration. In DC microgrids, the most frequently reported attacks on the secondary controller that affect both current sharing and voltage restoration are stealth cyber-attacks (Sahoo et al., 2019), denial of service (DOS) attacks (Liu et al., 2021), false data injection attacks (FDI) (Beg et al., 2017; Liu et al., 2021), and hijacking attacks (Sahoo et al., 2020).

In (Yu et al., 2023), a robust defense strategy against False Data Injection attacks is presented, which involves the implementation of a secure data transmission mechanism based on blockchain technology. This innovative approach is structured into four distinct phases. In the initial phase, the ith DC-DC converter uses a one-way hash function and a timestamp to encrypt a message containing the current value (I_i), which is then securely relayed to all converters (nodes) within the DC microgrid using peer-to-peer (P2P) communication. In the subsequent phase, each converter autonomously verifies the received messages. In the third phase, the verification results are disseminated across all network nodes through P2P communication. In the final stage, a voting mechanism is used to achieve a consensus on global validation.

However, while this blockchain-based scheme improves security, it introduces an inherent latency to the network. Each new current value undergoes a four-phase process that involves computationally intensive algorithms such as hashing and consensus mechanisms. To mitigate the latency associated with the blockchain and communication protocol, an advanced predictive secure distributed algorithm has been developed for secondary control.

However, it should be noted that the paper does not explicitly address the efficacy of this method in defending against alternative forms of cyberattacks. Furthermore, a comprehensive evaluation of the effectiveness of the scheme in safeguarding against potential attack vectors remains outstanding.

In their study (Dehghani et al., 2021) introduced a novel framework that combines Hilbert Hang Transform (HHT) with blockchain technology. This framework is designed to address the critical challenges of detecting False Data Injection (FDI) in a DC microgrid while ensuring the secure exchange of data among distributed generation agents, smart sensors, and loads. However, a noteworthy limitation of their approach lies in its omission of potential delays that may be introduced by the utilization of blockchain. Furthermore, it is important to highlight that the real-time nature of both FDI detection and secure data transmission through blockchain has not been thoroughly addressed. This has implications for the computational complexity and latency of the proposed system, which warrant further investigation.

6. Challenges and limitations of integrating blockchain technology on DC-microgrids

Blockchain and the Internet instill optimism for transforming the world and fostering a more democratic and equal society through their decentralized nature. However, lessons from the Internet's growth highlight that technology must be easily understandable for mass adoption. The current state of Blockchain for energy mirrors the early to mid-1990s Internet era, being obscure for most (Teufel et al., 2019).

Blockchain systems may face scalability issues when applied to microgrids with a large number of transactions. This can impact the system's ability to handle the increasing volume of data in a timely manner. Additionally, blockchain networks often consume significant amounts of energy, which may be a concern in the context of energy-efficient microgrids. The energy-intensive consensus mechanisms used in many blockchains can counteract the energy-saving benefits of DC microgrids. On the other hand, the confirmation times of transactions on blockchain networks can vary, leading to potential latency issues. In a DC microgrid where real-time responsiveness is crucial, delays in transaction validation may affect the overall performance of the system. Ensuring seamless integration between various components of a DC microgrid and the blockchain system can be complex. Standardization and interoperability protocols are essential to facilitate smooth communication between different technologies.

Moreover, the integration of blockchain into DC microgrids necessitates addressing numerous regulatory considerations. This involves adapting legal frameworks to accommodate smart contracts and addressing data privacy and security within energy trading. Ensuring compliance with existing energy regulations and standards while incorporating this innovative technology is crucial. Additionally, defining roles and responsibilities in the decentralized energy trading landscape is vital. Potential revisions or adaptations to existing laws to align with the implications of blockchain technology in the energy sector also require careful consideration. Achieving consensus among various stakeholders, including policymakers, utilities, and consumers, on regulating and governing these new systems is essential for successful implementation.

In addition, the initial setup costs for implementing blockchain in DC microgrids involve developing the necessary infrastructure, including hardware and software, and establishing the network. This often requires a significant upfront investment. Finally, integrating a new technology within an existing energy infrastructure might necessitate adjustments, further increasing initial expenses.

7. Summary

Blockchain technology, initially designed to secure cryptocurrency transactions, operates as a distributed ledger, securely linking records through cryptographic hash functions. Its widespread adoption in domains such as supply chain management, healthcare, intellectual property protection, energy trading, and the Internet of Things (IoT) underscores its versatility since inception.

In the context of DC microgrids, blockchain is still in its early stages of development, but has been proposed for use in peer-to-peer energy trading and to secure distributed control algorithms. However, there are challenges to using blockchain in secondary and tertiary control levels, such as the latency introduced by the computation algorithms, the scalability of the blockchain technology for larger DC microgrid, the energy consumed by the consensus mechanism, and the lack of standardization and regulation. These topics are active research areas.

References

Beg, O. A., Johnson, T. T. and Davoudi, A. 2017. Detection of false-data injection attacks in cyber-physical dc microgrids. IEEE Transactions on Industrial Informatics, 13(5): 2693–2703.

Bhutta, M. N. M., Khwaja, A. A., Nadeem, A., Ahmad, H. F., Khan, M. K., Hanif, M. A., Song, H., Alshamari, M. and Cao, Y. 2021. A survey on blockchain technology: Evolution, architecture and security. IEEE Access, 9: 61048–61073.

Dehghani, M., Niknam, T., Ghiasi, M., Bayati, N. and Savaghebi, M. 2021. Cyber-attack detection in dc microgrids based on deep machine learning and wavelet singular values approach. Electronics, 10(16). ISSN 2079-9292. URL https://www.mdpi.com/2079-9292/10/16/1914.

Hakak, S., Khan, G. W., Gilkar, B. Z., Assiri, A., Alazab, M., Bhattacharya, S. and Reddy, G. T. 2021. Recent advances in blockchain technology: A survey on applications and challenges. International Journal of Ad Hoc and Ubiquitous Computing, 38(1-3): 82–100. URL https://www.inderscienceonline.com/doi/abs/10.1504/IJAHUC.2021.119089.

Hashim, F., Shuaib, K. and Zaki, N. 2022, Oct. Sharding for scalable blockchain networks. SN Comput. Sci., 4(1). URL https://doi.org/10.1007/s42979-022-01435-z.

Khan, N. and State, R. 2020. Lightning network: A comparative review of transaction fees and data analysis. pp. 11–18. *In*: Prieto, J., Das, A. K., Ferretti, S., Pinto, A. and Corchado, J. M. (eds.). Blockchain and Applications. Cham, 2020. Springer International Publishing. ISBN 978-3-030-23813-1.

Koens, T. and Poll, E. 2019. Assessing interoperability solutions for distributed ledgers. Pervasive and Mobile Computing, 59: 101079. ISSN 1574-1192. URL https://www.sciencedirect.com/science/article/pii/S1574119218306266.

Liu, X. -K., Wen, C., Xu, Q. and Wang, Y. -W. 2021. Resilient control and analysis for dc microgrid system under dos and impulsive fdi attacks. IEEE Transactions on Smart Grid, 12(5): 3742–3754.

Mohanta, B. K., Jena, D., Panda, S. S. and Sobhanayak, S. 2019. Blockchain technology: A survey on applications and security privacy challenges. Internet of Things, 8: 100107. ISSN 2542-6605. URL https://www.sciencedirect.com/science/article/pii/S2542660518300702.

Musa, H. S., Krichen, M., Altun, A. A. and Ammi, M. 2023. Survey on blockchain-based data storage security for android mobile applications. Sensors, 23(21). ISSN 1424-8220. URL https://www.mdpi.com/1424-8220/23/21/8749.

Namane, S. and Ben Dhaou, I. 2022. Blockchain-based access control techniques for IoT applications. Electronics, 11(14). ISSN 2079-9292. URL https://www.mdpi.com/2079-9292/11/14/2225.

Negka, L. D. and Spathoulas, G. P. 2021. Blockchain state channels: A state of the art. IEEE Access, 9: 160277–160298.

Pee, S. J., Kang, E. S., Song, J. G. and Jang, J. W. 2019. Blockchain based smart energy trading platform using smart contract. In 2019 International Conference on Artificial Intelligence in Information and Communication (ICAIIC), pp. 322–325.

Sahoo, S., Mishra, S., Peng, J. C. -H. and Dragičević, T. 2019. A stealth cyber-attack detection strategy for dc microgrids. IEEE Transactions on Power Electronics, 34(8): 8162–8174.

Sahoo, S., Peng, J. C. -H., Mishra, S. and Dragičević, T. 2020. Distributed screening of hijacking attacks in dc microgrids. IEEE Transactions on Power Electronics, 35(7): 7574–7582.

Scheid, E. J., Hegnauer, T., Rodrigues, B. and Stiller, B. 2019. Bifröst: A modular blockchain interoperability api. In 2019 IEEE 44th Conference on Local Computer Networks (LCN), pp. 332–339.

Sun, X., Yu, F. R., Zhang, P., Sun, Z., Xie, W. and Peng, X. 2021. A survey on zero-knowledge proof in blockchain. IEEE Network, 35(4): 198–205.

Teufel, B., Sentic, A. and Barmet, M. 2019. Blockchain energy: Blockchain in future energy systems. Journal of Electronic Science and Technology, 17(4): 100011. ISSN 1674-862X. URL https://www.sciencedirect.com/science/article/pii/S1674862X20300057.

Yu, Y., Liu, G. -P. and Hu, W. 2023. Blockchain protocol-based secondary predictive secure control for voltage restoration and current sharing of dc microgrids. IEEE Transactions on Smart Grid, 14(3): 1763–1776.

Yuen, T. H., Sun, S. -F., Liu, J. K., Au, M. H., Esgin, M. F., Zhang, Q. and Gu. D. 2020. Ringct 3.0 for blockchain confidential transaction: Shorter size and stronger security. In Financial Cryptography and Data Security: 24th International Conference, FC 2020, Kota Kinabalu, Malaysia, February 10–14, 2020 Revised Selected Papers, pp. 464–483, Berlin, Heidelberg, 2020. Springer-Verlag. ISBN 978-3-030-51279-8.

Zeng, Z., Li, Y., Cao Y., Zhao, Y., Zhong, J., Sidorov, D. and Zeng, X. 2020. Blockchain technology for information security of the energy internet: Fundamentals, features, strategy and application. Energies, 13(4). ISSN 1996-1073. URL https://www.mdpi.com/1996-1073/13/4/881.

CHAPTER 4

Digital Twin Framework for Monitoring, Controlling and Diagnosis of Photovoltaic DC Microgrids

Ana Cabrera-Tobar[a],* and *Giovanni Spagnuolo*[b]

1. Introduction

The rapid advancement of technology, particularly the Internet of Things (IoT), has driven the transformation of Industry and Energy 4.0. IoT enables the interconnection of billions of devices, making applications smarter, faster, and more efficient. By 2025, it is expected that around 27 billion devices will be interconnected using IoT (Analytics, 2021). Integrating sensors, data collection, big data analytics, and artificial intelligence have revolutionized various industries, enabling the development of preventive maintenance, diagnostics, and enhanced automation features (Bazmohammadi et al., 2021).

In the energy sector, microgrids and smart cities are leveraging the power of IoT to create a communication network among various components, including power generators, weather stations for forecasting, loads, and end-users. This interconnected infrastructure enhances energy management, making it more efficient and sustainable (Zia et al., 2020). For instance, in smart grids, IoT plays a vital role in collecting real-time data from distributed energy resources such as solar panels, wind turbines, and energy storage systems. These data enable grid operators to monitor and optimize the performance of these resources, balance

[a] Polytechnic University of Milan, Dipartimento di Energia, Via Lambruschini 4, 20156 Milano, Italy.

[b] University of Salerno, Via Giovanni Paolo II, 132 - 84084 Fisciano (SA).

* Corresponding author: anakarina.cabrera@polimi.it

supply and demand, and enhance grid resilience. By incorporating IoT technologies, energy management systems can make informed decisions based on accurate data and predictive analytics, leading to more efficient energy distribution and reduced environmental impact (Tanyingyong et al., 2016).

Microgrids and smart cities are using IoT to communicate various components-power generators, the forecasting weather stations, the load, the user that are helping to enhance energy management to make it more efficient (Ali et al., 2021). The introduction of IoT in microgrids has not only led to the development of smarter Energy Management Systems but has also paved the way for a more dynamic, intelligent, and reliable system called Digital Twin (DT) (Hacı Bektaş et al., 2021).

NASA defines DT as mirror digital shadows communicating with their real flying systems through advanced sensors, enabling monitoring, predicting, and evaluating the system's state (Glaessgen and Stargel, 2012). In this context, DT provides continuous forecasts of the vehicle's health and remaining useful life to enhance mission success. Tao et al. define DT as digital models that simulate the behavior of a physical system in real-time, utilizing available data (Tao et al., 2019). The key components include the digital model, the physical part, and the data interconnection. However, recent definitions have expanded to include health diagnosis, prevention, maintenance, and data analytics for processing data from the physical and model layers. In this chapter, the definition of a Digital Twin is as follows:

Digital Twin is a living and adaptable model connected with its physical counterpart, where both continuously exchange data for prediction, management, control, and decision-making. It comprises four main components: the physical part, the adaptable digital model (digital shadow), real-time data connection, and prediction and analysis. Additionally, intelligent supervisory and control management plays a crucial role in leveraging the capabilities of DT.

DT has gained significant attention across aerospace, manufacturing, healthcare systems, agriculture, automotive, and robotics industries. For example, in the aerospace industry, DTs are employed to monitor and evaluate the performance of aircraft components, optimize maintenance schedules and enhance safety (Botín-Sanabria et al., 2022). In manufacturing, DT enables virtual simulations, predictive maintenance, and optimization of production processes (Kritzinger et al., 2018). Similarly, in healthcare, DT aid in patient monitoring, treatment planning, and personalized medicine (Bartsch et al., 2021).

In the context of energy systems, DT has garnered significant attention and is being increasingly applied to enhance the performance, efficiency, and reliability of various energy assets and infrastructure. According to recent studies, the global market for Digital Twin technology in the energy sector is projected to reach $1.3 billion by 2026 (Markets and Markets, 2020). This exponential growth underscores the transformative impact and increasing adoption of DT in the energy industry. DT provides valuable insights into the energy sector operation, enables predictive maintenance, and facilitates intelligent decision-making by creating a virtual replica of energy systems such as power plants, renewable energy sources, microgrids, and buildings. Companies such as General Electric and ABB have pioneered DT for power systems applications. For instance, General Electric developed DT for wind farms for operation and maintenance, increasing its efficiency by 20% in 2015.

In PV and DC (PVDC) microgrids, DT can control and diagnose the system's behavior. DT enables real-time monitoring, predictive analytics, and intelligent management by creating a virtual replica of the PV system, including solar panels, converters, storage systems, and loads. This empowers system operators and stakeholders to make informed decisions, optimize energy generation, minimize downtime, and improve the overall health and performance of the microgrid. In the case of PV applications, DT has just recently been applied. In 2020, Helios IoT Systems presented the first DT for a PV power plant to predict energy loss utilizing weather and electrical data from the power plant at various stages, including PV panels, DC and AC cables, transformers, and inverters. The result is the prediction of the energy losses, but no services are delivered, rather than just informative. Recent literature Jain et al. (2020) and Arafet and Berlanga (2021) show very few attempts of DTs for fault and anomaly analysis of only monofacials PV panels. A similar approach is presented in (Wunderlich and Santi, 2021) for only inverters. However, they fail to use the DT to provide any service to enhance the reliability of the PV generator. Additionally, the models presented do not adapt in real-time and do not consider the interaction among all the components involved, e.g. (Arafet and Berlanga, 2021). Other researchers have focused only on the communication layer but not on the modeling or the services (Yuan and Xie, 2023).

Despite the growing interest in DT, it is necessary to have a consensus on concepts, architecture, modeling techniques, and communication technology in the field of PV. The DT should rely on a holistic solution between the interaction among the various components so the PVDC microgrid can benefit from it. Thus, the present chapter aims to shed light on the capabilities and implications of Photovoltaic DT (PVDT), paving the way for a new energy control and management era. Through this chapter, we aim to provide a critical literature review to approach various topics regarding the DT framework for photovoltaics to discover research gaps and trends. We reviewed a total of 100 articles covering its main components, suitable modeling, control, and diagnosis techniques. The chapter is divided into six sections. Section 2 explains the DT framework for PVDC microgrids considering three main layers: the physical, the digital, and the communication. Then, the modeling techniques for PV in the context of DT are discussed in Section 3. A similar discussion is developed for DC-DC converters and their control in Section 4. In Section 5, we explain the services provided by a DT, taking into account the supervised control and management focused on preventive maintenance.

2. DT framework for photovoltaic DC microgrids

The architecture of DT for PV DC microgrids has three fundamental layers: the physical world, communication, and the digital world. A general framework about the components of the digital twin is illustrated in Figure 1.

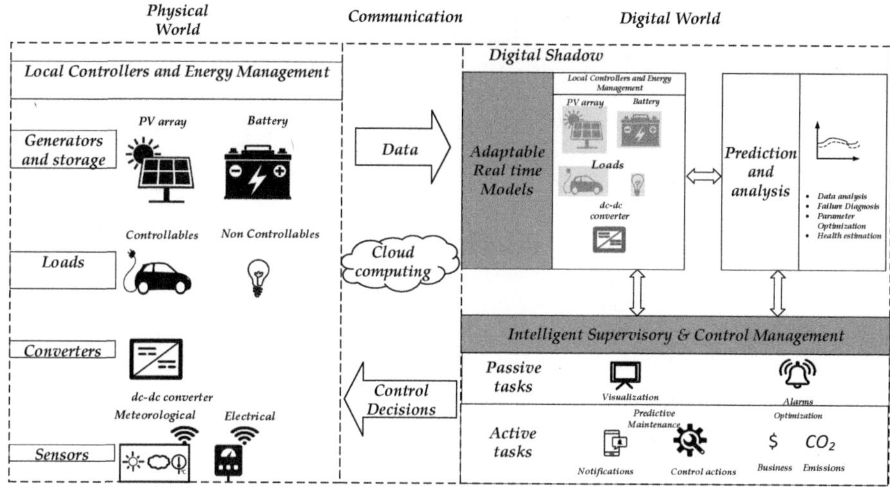

Figure 1. DT for PVDC microgrid: Physical world and digital world.

2.1. Physical world

The physical world represents the actual PV DC microgrid components and their operation. It encompasses PV panels, power converters, energy storage systems, loads, and their interconnections.

A PVDC microgrid can operate in isolated or non-isolated mode. In the case of an isolated PVDC microgrid, there is no connection to the main grid. In contrast, a non-isolated PV DC microgrid is connected to the grid (Planas et al., 2015). The main components of an isolated PVDC microgrid are interconnected through a common DC bus, utilizing power converters. Photovoltaic panels, batteries, and DC loads such as telecommunication antennas and electric vehicles (EVs) are connected to a DC/DC converter. However, if there are AC loads such as washing machines, motors, induction stoves, or electrical heaters, they are connected to a DC/AC converter. Additionally, multiple DC buses may exist within the microgrid, depending on the voltage magnitude and the number of components operating at specific voltage levels (Modu et al., 2023). This modular approach allows for greater flexibility and efficient utilization of resources.

In the case of a non-isolated PVDC microgrid, the DC bus converts the generated energy into AC using an inverter. The AC load is then connected to a single AC bus, enabling the utilization of standard AC-powered devices. It's

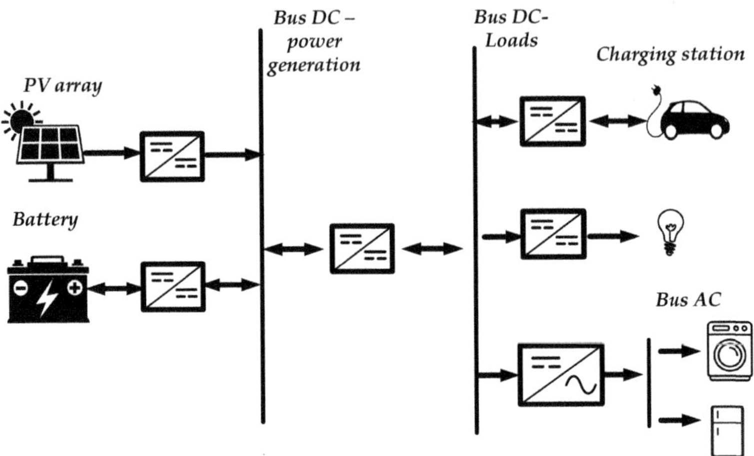

Figure 2. Generic PVDC microgrid topology.

worth noting that interconnecting the microgrid with the main grid may require a transformer to facilitate the necessary voltage levels and ensure compatibility (Cabrera-Tobar et al., 2016). Figure 2 provides a visual representation of a combined solution for a DC microgrid.

2.2. Communication

The communication layer of the digital twin for PV DC microgrids involves various technologies and protocols to facilitate efficient data exchange and control among different components (e.g., PV panels, sensors, and inverters). Several communication technologies have been widely adopted, such as RS485, Modbus, Bluetooth, and Ethernet-based protocols like Ethernet/IP and Profinet. These technologies enable reliable and real-time communication, allowing seamless integration and coordination among the microgrid components (Samanta et al., 2020). For instance, in a PV-based charging station located in Trieste, the communication used is a mix between Ethernet and RS485 (Cabrera-Tobar et al., 2022). The data coming from the sensors use RS485. Meanwhile, the communication with the inverter, the storage, and the human interface devices uses its own Ethernet network.

Additionally, IoT protocols like MQTT (Message Queuing Telemetry Transport), CoAP (Constrained Application Protocol), AMQP (Advanced Message Queuing Protocol), and XMPP (Extensible Messaging and Presence Protocol) are also integral to the operation of the digital twin in PV DC microgrids. MQTT facilitates secure, lightweight messaging among the components, ensuring efficient data exchange (Manowska et al., 2022). Similarly, CoAP provides a lightweight and reliable communication protocol based on the REST architectural style, enabling seamless interaction between devices and services in the microgrid (Tanyingyong et al., 2016). AMQP also provides a reliable and secure messaging protocol (Tanjimuddin et al., 2022). Meanwhile, XMPP is designed for fast and efficient messaging, although their lack of robust security features can pose potential risks to the operation of the microgrid.

In a study comparing MQTT, CoAP, AMQP, and XMPP for use in DC microgrids, three key parameters were evaluated: packet overhead, latency, and scalability (Kondoro et al., 2021).[1] CoAP exhibited 44% fewer bytes produced and faster data transfer, making it efficient for constrained environments. However, it showed poorer scalability than the other protocols, as it needs to establish security parameters for every data packet. In contrast, the other protocols do so at the beginning of the transmission.

The complex interaction among the various components of the PVDC microgrid, including measurements, decisions, and supervision, can make the communication layer vulnerable to cyberattacks. One type of attack affecting IoT devices is Distributed Denial of Service (DDoS), which aims to interrupt the flow of information between the devices. In the case of microgrids, this attack can affect the energy balance and cause voltage disruption (Hasan et al., 2023). Thus, the control of DC microgrids should be designed to be resilient to cyber-attacks. Emulators and observers could help to detect where the disruption of information occurs, helping the control algorithm to provide adequate power flow and voltage stability (Cabrera-Tobar et al., 2018; Sadabadi et al., 2022). On the other hand, the attacks can manifest as fake data, malicious software, and data leakage, posing a risk to the data privacy of users connected to the DC microgrid (Mustafa et al., 2019).

On the one hand, the DC microgrid control and the DT should acknowledge this vulnerability and incorporate it into their control mechanisms to enhance reliability in the presence of communication failures. For example, Sadabadi et al. proposes a resilient control mechanism for voltage and energy management to withstand cyberattacks in any part of the network (Sadabadi et al., 2022). On the other hand, it is necessary to enhance communication protocols by considering cryptographic methods that encrypt data to mitigate corruption at the source. However, this introduces computational complexity, which becomes a challenge for real-time operation. To address this, Yu et al. (2022) propose using blockchain and Byzantine ideas for a hierarchical control framework in a DC microgrid, reducing deficiencies in real-time communication.

The communication layer in the digital twin faces several challenges. First, there is a need for protocol standardization to facilitate implementation across different devices in DC microgrids. Second, security constraints must be incorporated to protect data transmission among components and in the digital realm. Third, user data confidentiality is a crucial aspect to consider. Fourth, real-time communication is essential for parameter and performance data and promptly

---

[1] Packet overhead refers to the amount of extra packet data due to the security extensions which can affect the performance of the communication in constrained environments. Latency refers to the time data travels from one component to another. Scalability refers to the capacity of the protocol to be used with a higher number of components and data.

communicating faults and emergency control actions, necessitating reduced latency.

2.3. *Digital world*

As a key component of the digital twin architecture, the digital world comprises several layers that enable advanced control, prediction, and analysis. At the core of the digital world is the adaptable modeling layer. This layer involves developing and implementing mathematical models and algorithms that accurately represent the behavior of the PVDC microgrid components. These models continuously adapt and update based on real-time data from sensors and IoT devices, ensuring the digital twin's alignment with the physical system.

The prediction and analysis layer leverages data analytics techniques to process and analyze the collected data from the physical system. This layer enables proactive decision-making and optimizing energy efficiency in the PV DC microgrid by detecting patterns, identifying anomalies, and performing predictive maintenance. It aids in identifying potential failures, provides insights for system performance improvement, and supports the implementation of preventive measures.

The supervisory control and management layer integrate the outputs from the prediction and analysis layer with advanced control algorithms. It encompasses real-time monitoring, fault detection, adaptive control, and intelligent decision-making based on the DT's insights. By continuously receiving and analyzing data from the physical system, this layer enables optimal operation, efficient energy management, and improved overall performance of the PVDC microgrid.

By adopting this architecture, the DT of a PVDC microgrid becomes a powerful tool for control and diagnosis. It establishes a dynamic connection between the physical and digital worlds, allowing continuous data exchange, prediction, analysis, and intelligent supervisory control. The DT enables enhanced monitoring, fault detection, and preventive maintenance, ultimately increasing system reliability, longevity, and optimized energy management.

Subsequent sections will delve deeper into the digital world and its various layers. The adaptable modeling, prediction, analysis, and supervisory control and management layers will be discussed in detail, showcasing their significance in leveraging the power of DT for control and diagnosis in PVDC microgrids.

3. PV modeling

The success of integrating PV array into DT hinges on accurate models that can automatically adapt to operational changes, aging, and material degradation. These models aim primarily to forecast output power and could aid in anomaly detection. Hence, the key questions are: what technique can ensure a reliable and adaptable model for the PV array?, and what parameters are crucial for the model, control, diagnosis, and prognosis? This section explores two types of PV modeling—physical and data-driven models.

3.1. Physical model

The output power of a PV solar cell is ideally determined by the solar irradiance (G) on its surface, by the cells operating temperature, and by the operating voltage, which has to be lower than the open circuit one. This relationship can be modeled using a single-diode model (SDM). However, to account for electrical losses, a more accurate model includes a series resistance (R_s) and a parallel resistance (R_p) (Figure 3). Alternatively, a model with two or three diodes can be employed to represent both the non-ideal behavior of the diode and the optical losses resulting from diffusion and recombination (Qais et al., 2022). Although the one-diode model is commonly used nowadays to depict the electrical characteristics of PV solar cells (Dolara et al., 2015), a three-diode model may be more beneficial when the digital twin requires representation of the optical aspects.

The SDM of a PV cell can involve three to five parameters essential for an accurate circuit representation. When using three parameters, only the diode ideal factor (A), the diode's inverse saturation current (i_o), and the short circuit current (i_{sc}) are considered, neglecting the electrical losses. On the other hand, the five-parameter model includes the additional consideration of series resistance R_s and parallel resistance R_p (Niccolai et al., 2018). The governing equations for the one-diode model with five parameters are elaborated in (Cabrera Tobar, 2018). Explicit equations and iterative solutions of nonlinear systems are commonly employed to determine the five parameters of different PV panels under non-standard test conditions. However, it is essential to note that parameter estimation errors may arise in either case (Piazza et al., 2017).

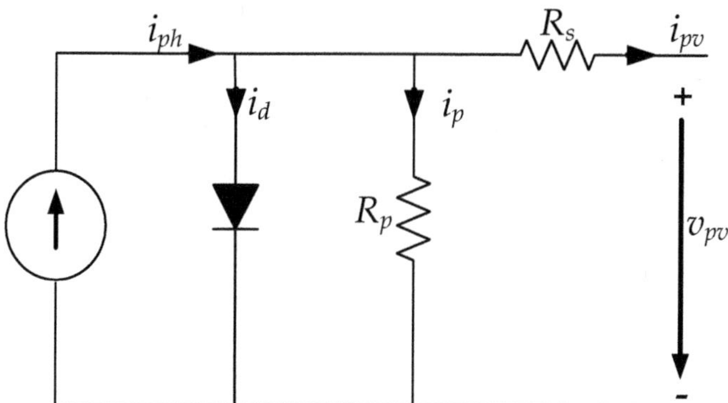

Figure 3. One diode-five parameter model of a PV solar cell. (Reprinted from Solar Energy Journal, Vol. 140, A. Cabrera-Tobar, E. Bullich-Massagué, M. Aragüés-Peñalba, O. Gomis-Bellmunt, "Capability Curve Analysis of photovoltaic generation systems", Pages 255–264, Copyright (2023), with permission from Elsevier).

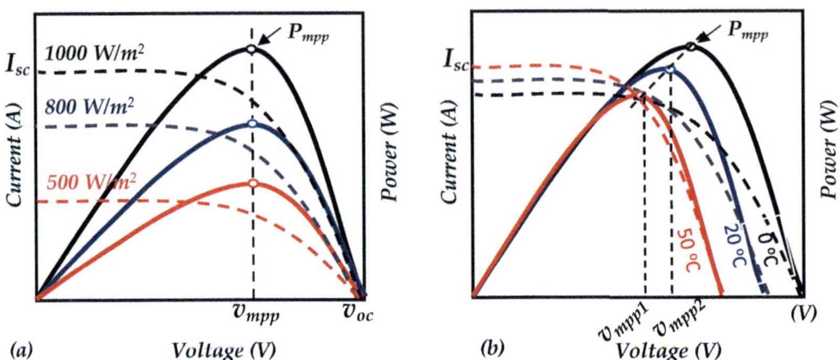

Figure 4. I-V and P-V curves of PV solar cell (a) Variable solar Irradiance and $T_a = 25^{\circ}C$ and (b) Variable ambient temperature and $G = 1000W/m^2$. (Reprinted from Solar Energy Journal, Vol. 140, A. Cabrera-Tobar, E. Bullich-Massagué, M. Aragüés-Peñalba, O. Gomis-Bellmunt, "Capability Curve Analysis of photovoltaic generation systems", Pages 255–264, Copyright (2023), with permission from Elsevier).

The I-V and P-V curves can describe the PV array model's performance (Figure 4). On these curves, four values are important: the I_{sc}, the maximum power point (P_{mpp}), the open circuit voltage (v_{oc}), and the voltage to get the maximum power point at every solar irradiance (v_{mpp}). It is important to mention that the PV solar cell's power production is also influenced by the ambient temperature. The solar cell temperature impacts the values of (v_{oc}) and v_{mpp}, consequently directly affecting the power operating point. While the gradient effect due to temperature is typically mentioned in data sheets, the thermal equation employed for modeling is often static and used to derive the I-V and P-V curves rather than for dynamic modeling purposes.

In literature, due to the importance of estimating correctly the T_c, several models have been presented. Santos et al. (2022) discuss 33 correlations found in the literature, with key parameters including solar absorbance, electrical efficiency, solar irradiance, ambient temperature (T_a) and wind speed. The most widely used thermal model for solar cells typically includes ambient temperature(T_a), cell temperature (T_c), solar irradiance (G), Normal Operating Cell Temperature ($NOCT$), and a reference solar irradiance (commonly set at 800 W/m^2) (Markvart, 2012). The equation for this model is expressed as follows:

$$T_c = T_a + G \times \frac{NOCT - 20}{800} \tag{1}$$

The main drawback of this equation is that it does not account for the heat transfer coefficient between the solar cell and its surroundings, which includes factors like wind, mounting structure, and specific installation conditions. Furthermore, the equation lacks dynamic considerations (Migliorini et al., 2017). The parameter NOCT is typically assumed to be between 40 and 45°C for monocrystalline and polycrystalline modules with adequately ventilated mounting structures. However, new applications of PVDC microgrids, such as vertical installations in buildings, floating installations, and their use in agrivoltaics,[2] may have different NOCT values, which will be challenging to identify.

The influence of solar cell temperature on voltage is typically described by a thermal coefficient, which quantifies the variation of V_{oc} with respect to T_c. Although data sheets often provide this coefficient, estimation error can be associated with the real operation (Mihaylov et al., 2016). Furthermore, these coefficients can change over time due to solar panel degradation, making it necessary to estimate them for accurate solar panel modeling. Piliougine et al. (2021) present a practical guideline for estimating these coefficients using I-V curves obtained through a tracer method. This allows for subsequent determination of the electrical parameters within a specific and controlled temperature range. One drawback of this technique is that the PV panel needs to be temporarily out of operation to estimate the new coefficients.

In DT, it is essential to have models that can adapt over time and accurately represent the performance of PV panels. Real-time operation is also a requirement. Hence, techniques that offer reduced computational time and memory usage are necessary. Siddiqui and Arif (2013) present a comprehensive 3D model that considers the physical, electrical, and thermal aspects of PV panels. This model provides accurate predictions of PV power output and estimates the remaining useful life based on thermal and physical performance. However, due to the extensive computational time required, 3D modeling is unsuitable for DT implementation.

Physical PV models are utilized to estimate the power production of PV panels by considering their electrical and thermal characteristics. However, these models are prone to errors due to the dynamic nature of the parameters involved. These parameters can vary over time and are influenced by factors such as ambient conditions, operational changes, faults, shadowing, material degradation, and dust accumulation, among others (Dolara et al., 2015). As a result, relying solely on physical models may not be sufficient for DT applications, especially in situations where specific factors like shading play a significant role. In such cases, data-driven models can complement the physical models to provide a more realistic model.

3.2. *Data-driven models*

Data-driven models play a crucial role in the development of DT in various industries like automation, aerospace, manufacturing, and energy. Also, it is crucial in the modeling of Photovoltaics. The data-driven models use available data to

---

[2]Agrivoltaics is the use of PV generators in farms to provide sustainable energy. Its installation should not affect the use of land for agricultural activities. Thus, vertical installations are commonly employed. For this application, bi-facial PV panels are preferred as they can receive solar irradiance for both sides of the PV panels.

create a mathematical model aiming to simulate and predict the performance of the real application. In contrast with physical models, data-driven models can offer various advantages: (i) accurate representation, (ii) adaptability and flexibility, (iii) real-time operation, and (iv) forecasting and diagnosis. However, some data-driven models (e.g., Deep Learning (DL), Reinforcement Learning (RL)) require a large quantity of data for the specific application, high quality of data, and significant computational resources (Aslam et al., 2021).

Real-time modeling is a key aspect of DT as it has to learn and adapt to prioritizing computational resources actively. The main techniques used are: (i) Online Machine Learning (OMSL), (ii) Recurrent Neural Network (RNN), (iii) Kalman Filtering, (iv) Particle Filters, (v) Online adaptive models, (vi) Data stream mining, (vii) Ensemble learning, and (viii) Transfer Learning (Alexopoulos et al., 2020). In the field of photovoltaics, there are some specific challenges that real-time data-driven models have to consider. They have to handle non-linearities, weather variability, rapid adaptation due to operational changes, and scalability.[3] Here, we analyze the main data-driven techniques for real-time modeling and their applicability in the field of photovoltaics:

Online Machine Learning: OMSL uses algorithms to update the model as new data arrives, enabling fast learning and prediction times. One of the techniques inside this category is Online Sequential Extreme Learning Machines which offers a low computational overhead and upgrades the training data sets block by block in real-time, making it suitable for real-time application with reduced forecast error (Zhang et al., 2018). However, it is not suitable for non-linear systems, as in the case of photovoltaics. Nevertheless, Behera et al. (2018) used this technique to model the PV power production where the weights were updated using Particle swarm optimization (reducing the non-linearity), showing a good performance regarding computation time and error. However, the same study showed that this algorithm lacks scalability, as the computational complexity will increase if the operation characteristics change.

Recurrent Neural Network: It is suitable for sequential data with temporal dependencies as time-series variables. Mellit et al. (2020) compare various types of RNN for PV power production, e.g., Long Short-Term Memory (LSTM), Bidirectional LSTM (BiLSTM), Gated Recurrent Unit (GRU), Bidirectional GRU (BiGRU), and One-Dimension Convolutional Neural Network for various time steps with one-step and multiple steps ahead. The study proved that GRU and LSTM provided a very good accuracy for a one-minute forecast and one step ahead. However, the accuracy of the data-driven model reduces due to cloudy days, the quantity of training data for normal and abnormal operations, and the degradation of the PV panel. Moreover, RNN techniques can suffer

---

[3] Scalability occurs when the systems scale up in size and complexity (e.g., One more PV string is added), increasing the volume of data and computational requirement.

from longer training times, making it non-suitable for real-time modeling nor for applications where no historical data is available, e.g., new PV installations.

Kalman Filtering: It provides real-time estimation considering measurements, uncertainties, and the system's dynamics. In contrast with the previous technique, it does not use historical data. Hence, it can effectively handle abrupt changes due to the weather conditions causing abnormal operations. However, it can cause estimation errors as it cannot handle non-linearity or non-Gaussian systems. Nevertheless, Yang et al. (2021) used Kalman-Filtering for the PV power prediction, dynamically correcting the model using real-time measurements. But it has an 18% chance of error when a drastic weather variation occurs. Thus, a dynamical tuning technique of the Kalman Filter's main parameters could help reduce its error, as explained by Soubdhan et al. (2016). Finally, this technique could be computationally expensive for large-scale systems. Thus, its scalability for real-time modeling is limited.

Particle Filters: They are used for state estimation of non-linear and no-Gaussian systems. They operate using three steps: predicting, updating, and resampling using real-time measurements (Zhang et al., 2023). They are based on sequential Monte Carlo methods. In PV, it is commonly applied to estimate the PV system's state, including solar irradiance, dust, temperature, and electrical characteristics under variable conditions (see Gao et al., 2022). Particle Filters are inadequate for large systems as they affect accuracy and computational complexity. Also, it needs careful tuning of the parameters, which can be a drawback in PV systems.

Online Adaptive Models: These models adapt to changing data by continuously updating their parameters and capturing system dynamics. Such models can be Adaptive neural networks and adaptive regression models. As they continuously learn, they offer flexibility when the system's size increases and there is a lack of historical data. For instance, Massidda and Marrocu (2017) used an adaptive regression model to forecast the PV power of a new PV power plant in Germany. The results showed low errors despite the low number of training samples and features included in the model. However, the online adaptive model is sensitive to noise and can be unstable when quick adaptation is required.

Data Stream Mining: This technique is suitable for real-time modeling, as it can handle high-speed and continuous data streams. It adapts the model according to the new data distribution and can detect anomalies. Some techniques inside this category are Online Clustering and Stream Regression (Kumar and Singh, 2017). For example, online clustering has been used for fault detection in PV. Because the ambient conditions affect the PV data, the cluster can have arbitrary shapes for each data stream. If a fault occurs, the new cluster will differ from the previous one saved in normal operation, and thus the fault can be detected (Cai et al., 2020). It is essential to note that this technique will handle large quantities of data. Thus, the computational resources could limit the performance of this method. Moreover, in PV applications, the noise in the inherent data could impose challenges in this high-speed methodology as it is required to preprocess the data. Due to the nature of variation of the PV parameters in time, it can cause a change in the underlying patterns, known as Concept Drift. Thus, the data stream mining could require updating the relationships between the main parameters over time (Lu et al., 2019).

Transfer Learning: Transfer Learning (TL) aims to develop adaptable neural networks (NN) for digital twin (DT) applications, even with limited data (Alexopoulos et al., 2020). TL leverages past knowledge to enable online learning and solve new tasks, making this technique suitable for quick model deployment despite changing conditions. For example, Sarmas et al. (2022) proposed a TL-based LSTM model, demonstrating improved forecasting accuracy with only three months of data. Another recent study developed by Schreiber and Sick (2022) combines TL with Bayesian regression, enabling accurate PV power production forecasts with only seven days of data.

The comparison of the various techniques mentioned in this section is in Tables 1 and 2. RNN, TL, and Online Adaptive Models stand out regarding scalability and computational resource requirements. These techniques excel in managing non-linearities and can adapt to changing conditions. They also offer flexibility in working with or without historical data. Additionally, Particle Filters and Kalman Filters are effective in handling uncertainties. Efficiency in computational time is crucial for real-time modeling using data-driven techniques. Online machine learning algorithms like Online Sequential Extreme Learning Machines (OS-ELM) offer fast learning and prediction times, making them suitable for real-time applications with low computational overhead. However, more complex models like RNN may require additional computational resources and longer training times. Nevertheless, advancements in hardware acceleration (e.g., GPUs, specialized chips) and optimization algorithms have helped reduce the

Table 1. Comparison of data-driven techniques for real-time modeling in Photovoltaic Digital Twin (PVDT) applications (Part 1).

Technique	Hist. Data	Adaptability	Uncertainty
OMSL	Yes	High	Medium
RNN	Yes	High	High
Kalman Filter	Yes	High	Medium
Particle Filters	No	High	High
Online Adaptive Models	No	High	Medium
Data Stream Mining	No	High	High
TL	Yes	High	High

Table 2. Comparison of data-driven techniques for real-time modeling in Photovoltaic Digital Twin (PVDT) applications (Part 2).

Technique	Scalability	Comp. Res.	Non-linearities
OMSL	Medium	Low	Low
RNN	Medium	High	High
Kalman Filter	Low	Low	Medium
Particle Filters	Medium	Medium	High
Online Adaptive Models	High	Low	Medium
Data Stream Mining	High	Medium	Medium
TL	High	High	High

computational burden, enabling faster training and inference times for these models.

4. The role of the converter

The DC-DC converter plays a vital role in DC microgrids by stepping up or down the DC voltage. One of the key functions of the DC-DC converter is active power control within the microgrid. It regulates active power flow between different components, facilitating optimal power-sharing and distribution by adjusting the converter's duty cycle. Moreover, the DC-DC converter maintains the desired DC voltage level within the microgrid. The converter regulates the voltage through adjustments in the duty cycle, ensuring compatibility with the DC bus. To enhance control and modeling accuracy, observers and estimators have been developed. These techniques employ sensors and advanced algorithms to estimate system parameters in real-time, providing the control system with accurate information about the converter's operating conditions. This allows for better control strategies and fault detection. Artificial Intelligence (AI) techniques have recently been applied to power electronics and control systems, offering opportunities to improve further DC-DC converter performance, fault diagnosis, and energy management. In this section, we discuss the converter's role from the active power's point of view, and later we discuss the main modeling techniques suitable for DT.

4.1. Active power control

Depending on the specific application, the voltage rate, and the power, the DC-DC converter will have a different topology. Generally speaking, the converter is a matrix of switching devices, parasitic resistance due to the cables, a capacitor, and an inductance, and it may have a high-frequency transformer to step up the voltage and offer galvanic isolation. In a PV system, the converter has the PV's current and DC voltages as inputs. The converter's inputs and switching state

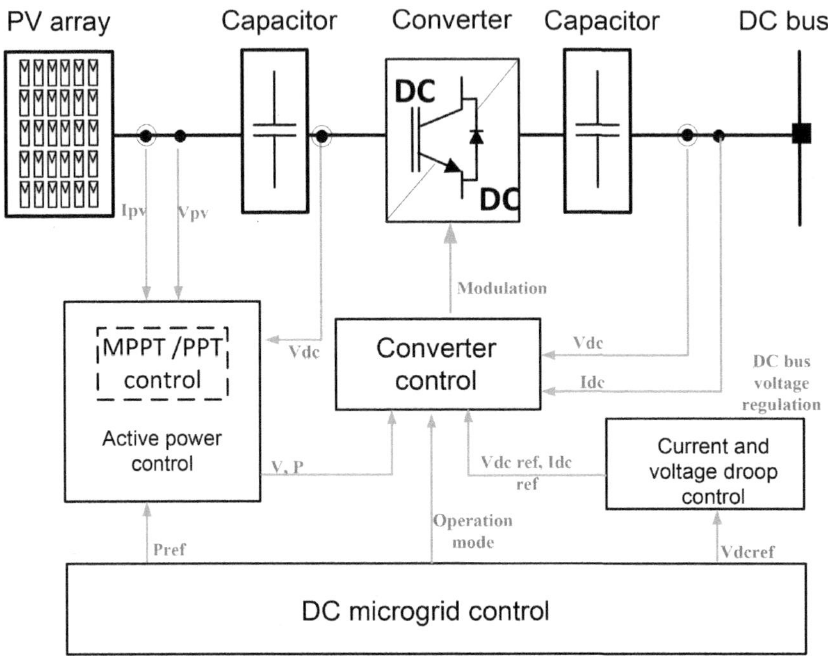

Figure 5. Control architecture of a PV DC-DC converter.

are fed into the controller (microprocessor, FPGA, or analog-based controller) to modulate the switches to manage the output power and the DC voltage.

In a PVDC microgrid, there are two control layers. The upper layer is ruled by the microgrid controller, which keeps the DC voltage at certain levels and avoids power imbalance. The second layer is the individual converter's control which controls the active power and the DC voltage, taking as reference the values sent by the upper layer (P_{ref} and V_{ref}) (Figure 5).

In the control of converters in DC microgrids, the controller typically incorporates feedforward connections from both the input and output to achieve the desired power and voltage operating points. When it comes to active power control, there are two common approaches: Maximum Power Point Tracking and Reference Power Point Tracking.

Maximum Power Point Tracking (MPPT): The MPPT control strategy aims to track the maximum power point of the PV system at any given instant. It continuously adjusts the operating point of the PV system to maximize the power output by considering factors such as solar irradiance and temperature. MPPT algorithms, such as Perturb and Observe (P&O) or Incremental Conductance (Beriber and Talha, 2013; Femia et al., 2012). Other conventional algorithms are: constant voltage, temperature gradient, short-circuit current and open circuit voltage method, global MPPT, fitting curve, scanning I-V curve, lookup table, and others (see Baba et al., 2020). However, these algorithms have the drawback of getting trapped on local MPP (LMPP) and have problems finding the global MPP (GMPP).[4] In addition, these algorithms have the drawback of low adaptability, thus unsuitable for DT.

Thus, artificial intelligence techniques are being used to track the GMPP, considering the non-linear characteristics of the P-V curve. Moreover, they can adapt in time and can provide a fast response during transient periods of time. Some techniques are fuzzy logic control (FLC), artificial neural network (ANN), genetic algorithm (GA), and Cuckoo search, among others (Bendib et al., 2014; Borni et al., 2017; Mosaad et al., 2019; Nguimfack-Ndongmo et al., 2022). However, the main drawback of these techniques in DT is the computing resource needs (see Yap et al., 2020). In Table 3, these techniques are compared considering: accuracy, speed, computing resources, and adaptability.

Table 3. Comparison of AI-based and Conventional heuristic-based MPPT techniques (L: Low, M: Medium, H: High).

MPPT Technique	Accuracy	Speed	Computing Resources	Adaptability
Fuzzy Logic Control (FLC)	M	H	L/M	H
Artificial Neural Network (ANN)	H	M	H	H
Differential Evolution (DE)	M	L	M/H	H
Genetic Algorithm (GA)	M	M	M/H	H
Particle Swarm Optimization (PSO)	H	M	M/H	H
Cuckoo Search (CS)	M	M	L/M	H
Firefly Algorithm (FA)	M	M	L/M	H
Hybrid Algorithms	H	M	M/H	H
Perturb and Observe (P&O)	L	H	L	L
Incremental Conductance (IC)	M	H	L/M	L/M
Hill Climbing (HC)	L	H	L	L
Constant Voltage	L	H	L	L
Fractional Short-Circuit Current	L	H	L	L
Fractional Open-Circuit Voltage	L	H	L	L
Scanning-Tracking of I-V Curve	M	M	M	L/M
Global MPPT (GMPPT) Segmentation Searching	L	H	L	L
Extremum Seeking Control	M	H	L/M	L/M

Reference Power Point Tracking (RPPT): In contrast to MPPT, the RPPT control strategy tracks a reference power set point determined based on various factors 6) (Zanatta et al., 2017), including:

- Energy Balance: The RPPT controller ensures that the power generated by the PV system matches the power consumed by the load or energy storage system within the microgrid. It maintains the overall energy balance by adjusting the converter's operation to match the required power levels.
- Power Curtailment: Sometimes, limiting the power produced by the PV system to a specific reference power agreed upon for the microgrid is necessary. This power curtailment can occur when the PV system generates more power than the microgrid can accommodate. The converter adjusts

---

[4]Shadowing conditions in the PV module or PV array provoke multiple MPP called local MPP. GMPP is the optimal MPP for the PV modules or array considering the combined effects of shading (Piliougine et al., 2022).

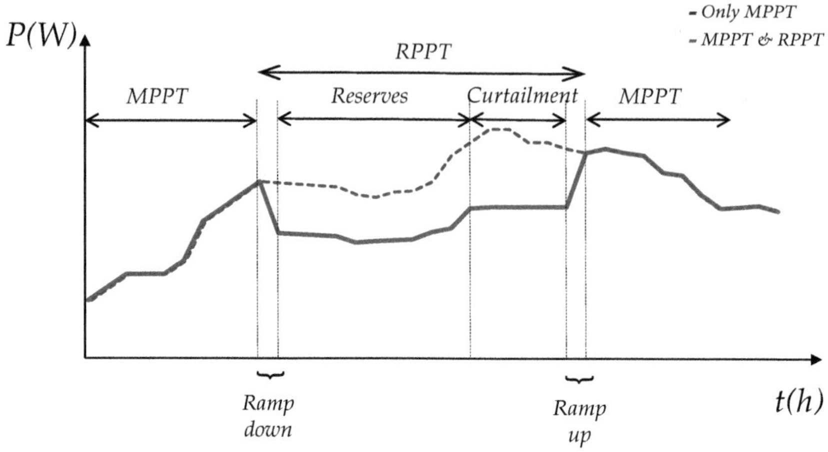

Figure 6. Active power control modes in a DC-DC converter for PV applications.

its operation to limit the power output to the agreed reference power level (Cabrera-Tobar et al., 2019).

- Ramp Rate Control: Rapid changes in solar irradiance can cause abrupt power variations in the PV system, which may impact the stability and reliability of the microgrid. The ramp rate control function in the RPPT controller helps the PV system increase or decrease its power output gradually, within a specified rate, to avoid potential damage to the load. This control strategy ensures smooth power transitions and prevents undesirable voltage fluctuations in the microgrid (Cabrera Tobar, 2018; Sukumar et al., 2018).

By employing RPPT control strategies, converters in DC microgrids can ensure that the power generation from PV systems aligns with the energy balance requirements, prevents power overflows through curtailment, and manages ramp rates to maintain stability during changing irradiance conditions. Cabrera-Tobar et al. (2019) provide insight into active power control considering these strategies.

4.2. Modeling

For DT, the modeling of the power converter has four main objectives: (i) to detect anomalies in operation due to thermal stress, (ii) to prevent electrical failures, (iii) to develop preventive maintenance, and (iv) to increase reliability. Thus, the converter's modeling should indicate the real counterpart's thermal and electrical characteristics, using variables such as input and output voltages and currents and temperature measurements. Moreover, the digital twin should cover the physical part, the converter's control, and switching performance. This section analyzes the various modeling techniques that could be used in DT, considering computational time and sources, real-time operation, and scalability. We consider electrical and thermal models, then black-box models, probabilistic twinning, and finalize with observers and estimators.

Electrical and Thermal Models: The modeling techniques for the electrical performance of the converters can be divided into: (i) average-value modeling, (ii) state-space modeling, (iii) switched mode modeling, and (iv) switched linear modeling (Wunderlich and Santi, 2021) (Frances et al., 2018). In the case of thermal models for DT, the most suitable are: (i) Lumped Thermal and (ii) variations of Finite Element Analysis (see Cavallaro et al., 2019) -Reduced Order Models, Proper Orthogonal Decomposition (see Haider et al., 2010), Model Order Reduction (see Nahvi et al., 2017). The main characteristics of these techniques are summarized in Table 4. The review paper, presented by Frances et al. (2018), discusses various averaging techniques as they offer fast computation and the representation of the dynamics could be improved using non-conventional methods such as Krylov-Bogoliuv-Mitropolsky or also floquet-based and cyclic averaging for rapid analysis.

Black-box Modeling Techniques: In DC microgrids and DC power converters, the electrical and thermal models' accuracy is limited to the manufacturer's data, which is not always available. In this case, black-box modeling represents the relationship between the input-output without considering the internal dynamics of the converter or its physical components. The main techniques in this category are Hammerstein-Wiener Modeling (HWM) and Data-driven modeling. HWM combines linear dynamic systems and static non-linearities. In DC-DC converters connected to a PV array, the static nonlinearity block could represent the voltage-current relationship of the PV array. Meanwhile, the dynamic linear block could represent the linear dynamics of the DC-DC converter. It can help with electrical or thermal modeling and adapt in real-time. However, it may require high-quality data to identify the model and to estimate the main parameters (Wunderlich and Santi, 2021). On the other hand, data-driven modeling can be used as a black box. However, the techniques should have the main characteristic of self-adapting. As discussed, online learning, Kalman, and particle filtering can be applied.

On the other hand, data-driven modeling can also be used for fault detection and diagnosis. Rouzbehi et al. (2019) use a local model network (LSM) to identify the converter dynamics with measured data considering variable modulation, switching frequency, input DC voltage, and controlled load. However, modeling its response under real data input from photovoltaics in data-driven models for DC-DC converters is also necessary. The current variation from the photovoltaic array and the MPPT control is essential in the modeling. For instance, Bindi et al. (2022) utilize a Multilayer neural network with Multi-Valued Neurons to model the DC-DC converter to determine faults using solar irradiance and ambient temperature as input variables. Table 5 compares black-box techniques regarding the

Table 4. Comparison of electrical and thermal modeling techniques for DC-DC converter suitable for real-time and DT applications (E: Electrical model, T: Thermal model, L: Low, M: Moderate).

Modeling Technique	Type	Comp. Time	Comp. Resources	Advantages	Disadvantages
Average-Value Modeling	E	L	M	Fast computation	Simplified representation of dynamics
State-Space Modeling	E	L	M	Accurate representation of dynamics	More complex implementation compared to average-value modeling
Switched Mode Modeling	E	M	M	Captures switching behavior and nonlinear effects	Requires detailed knowledge of converter operation
Switched Linear Modeling	E	M	M	Simplified representation of switching converters	Limited accuracy for nonlinear and high-frequency effects
Lumped Thermal Modeling	T	M	M	Fast computation	Limited accuracy for spatial temperature variations
Reduced Order Models	T	M	M	Fast computation	Reduced accuracy compared to full FEA simulations
Proper Orthogonal Decomposition (POD)	T	M	M	Reduced computational complexity	Limited accuracy for highly nonlinear systems
Model Order Reduction (MOR)	T	M	M	Reduced computational complexity	Limited accuracy for highly nonlinear systems

accuracy, computational efficiency, model adaptability in real-time, data requirements, and scalability.

Despite the importance of applying machine learning for thermal modeling in power converters, the literature is limited on this topic. Thermal modeling concentrates not only on one device but also on compact packaging with multiple semiconductors. Thus, the thermal performance of one power device is also affected by its neighbor. The thermal performance measurement and analysis take considerable time due to the interaction of the various power devices. This interaction is known as thermal cross-coupling, which is commonly neglected. Zhan et al. present an ANN to model this interaction with limited data using the thermal and cooling curves of the devices. The use of this model only takes

Table 5. Comparison of black-box modeling techniques for DC-DC converters.

Technique	Accuracy	Comp. Efficiency	Adaptability	Data Req.	Scalability
Hammerstein-Wiener Model	H	M	M	L-M	M
Online Learning	H	M	H	H	H
Kalman Filtering	H	H	H	H	H
Particle Filtering	H	L-M	H	H	M-H
Deep Neural Networks	H	L-M	L-M	H	M-H
Multi-Valued Neural Network	H	L-M	L-M	H	M-H
Support Vector Machine	M-H	H	L-M	H	M-H
Local Model Network	H	M	H	M-H	M-H

100 seconds and with an error of less than 1°C. These results are promising for including thermal modeling of power converters for real-time simulation (Zhang et al., 2020).

Probabilistic Twinning: According to Milton M. et al., probabilistic twining uses a probabilistic simulation model with random variables and compares it with its physical part to assess if the converter is operating inside its probable behavior (Milton et al., 2020). Probabilistic modeling helps to model the system considering uncertainties (stochastic parameters) that affect the converter's operation. These uncertainties can be peak current, manufacturing defects, and other disturbances. The probabilistic modeling could be time-consuming, but its application in a Field Programmable Gate Array (FPGA) could be suitable for real-time simulation for small converters.

Observers: The main purpose of observers is to estimate parameters and the system state, taking into account the input and output measurements from the real application. They can be used to detect anomalies by monitoring the system's behavior or to estimate internal variables that are not measured. In DT, this technique could help to predict, control and optimize the operation of the PVDC. For example, observers can be built to model the behavior of the power converter. It could focus on electrical or thermal parameters to provide a relationship between the input and output variables (see Cabrera-Tobar et al., 2018). Stella et al. (2018) create an observer based on a linear state space equation. The input of the steady state model is the solar irradiance, ambient temperature, and load's current. The output is the DC voltage at the terminals of the converter. However, it does not consider the thermal model of the converter and its effect on the operation. Andresen et al. (2016), developed an observer using as input the load current, the DC voltage, the state of the converter, and the case temperature. The model relates the junction temperature with the power losses, the thermal capacitance, and the thermal resistance. Although this technique has good results, the parameters of thermal capacitance and thermal resistance could change over time, and they should also be estimated and updated.

5. Supervisory control and management

Incorporating a DT into a supervisory control and data acquisition (SCADA) system is crucial for enhancing the system's decision-making capabilities. By analyzing historical data and current performance and predicting the system's state of health, the DT can effectively optimize energy management and utilization and maintain a PV system's healthy operation. The main layers of a SCADA are: (i) field layer (sensors), (ii) data acquisition and data storage, (iii) real-time monitoring, (iv) plant control layer, and (v) executive control layer (Hu Guozhen et al., 2009) (Figure 7).

Layers	SCADA	Digital Twin
Executive	Schedule Maintenance • Uses recommendation provider	• Preventive Control to reduce Maintenance • Schedule Maintenance using the health diagnosis
Controller	• Energy Management System • Optimization	• Energy Management System • Optimization considering also the preventive control
Real time Monitoring	• Human-Machine Interface • Alarms and notifications	• Human-Machine Interface • Alarms and notifications • Augmented Reality • Possibility to see the real state of the elements: e.g. PV solar cell by layers
Prediction	• Not developed	• Fault and health diagnosis • Parameter variation • Remaining Useful Life
Data acquisition	• Data storage • Data Transfer	• Data storage • Data Transfer • Big data analysis
Field layer	• Sensors and devices: electrical and thermal measurements, weather stations, converters, PV solar panels, batteries	

Figure 7. Architecture layers of a SCADA with and without a DT.

In the field layer, we find all the sensors and actuators, e.g., breakers, wind speed sensors, voltage, and current sensors. The second layer is the data acquisition and the data storage from the field, which can be stored in the cloud, and the

DT can access it anytime. The control layer is the common management of the PVDC, considering energy management, energy optimization, and keeping the normal operation of the PVDC. The next layer is real-time monitoring, where a Human-Machine Interface is commonly used to monitor the main parameters and state of the PVDC. For instance, it can show the battery's state, the PV power, and the power consumption. This interface depends on the application and the amount of data available to the user. In contrast, an augmented reality interface could be added in the DT to see the electrical and thermal performance of the various devices (Pargmann et al., 2018). Executive management is commonly added so that the operator can act in case of an emergency, or for maintenance purposes, like the disconnection of the load. However, the executive management with the DT could be enhanced by providing prognostics and health management (PHM). This service can provide real-time and predictive maintenance of the PV array and the DC-DC converter. It can have three layers: (i) fault detection and diagnosis, (ii) remaining useful life analysis, and (iii) control actions.

5.1. Fault detection and diagnosis

DT has been developed to provide a prognosis of the device's lifetime, its operation characteristics, and its sensitivity to failure. Thus, the DT's controller must constantly compare the physical part with its digital counterpart. Any differences regarding voltage, current, and power could indicate variations of the digital model due to operation and aging or a failure in some components. Thus, after the digital model is created, an analysis of the parameters has to be developed to detect the reason for any anomaly. The correct fault diagnosis can enhance the system's operation by tracking down the cause and clarifying the fault as soon as possible, which helps reduce maintenance in the long run.

Conditioning monitoring, fault diagnosis, and fault detection are the main aspects of failure management. In the case of photovoltaics, the fault diagnosis of the PV system is commonly developed in two sections, one considering only the PV panel and the other just the converter. In the case of PV panels, failures may occur due to soiling, shading conditions, cell breakage, bypass diode failure, interconnection faults, internal faults, hot spots, and degradation (Chine et al., 2016). In the converter, the failures could occur due to the passive elements such as the capacitor and the inductor or the state of the power converter devices or related to the controller and its sensors. For instance, Leva et al. (2019) present a generic PV module fault diagnosis using micro-inverters and comparing the real output power with the forecasted one. Thus, a deviation from the measured and forecasted values is evaluated in real-time and offline for further inspection. However, they do not detect the reason for the fault nor the effect of the microinverter used.

Preprint et al. (2021) studied in detail the fault diagnosis and detection for the DC-DC converter. The authors apply the following data-driven techniques: (i) principal component analysis, (ii) multilinear principal component analysis, (iii)

uncorrelated multilinear principal component analysis, (iv) Fast Fourier Transformation preprocessing-based multilinear principal component analysis, and
(v) uncorrelated multilinear principal component analysis. The faults diagnosed
were related to the inductance and capacitor failures detected by the voltage and
current values variation.

However, in Digital Twin, it is necessary to have a holistic point of view between the PV panel and the converter. In the case of ac microgrids, Jain et al.
(2020) studied the fault diagnosis of PV systems (PV panel and converter). The
authors propose a methodology to detect ten different types of faults by evaluating a residual error vector evaluated between the estimated and the measured
output. However, the main drawback of the proposed technique is that the faults
are detected considering a specific reference value of the current at the various
points of the PV system, but there is a lack of analysis if these references will
change in time due to operation or the effect of aging.

5.2. *Remaining useful life*

The remaining useful life (RUL) is a technique to estimate the number of years
with a specific component to function according to the operating parameters
given by the manufacturer, the monitored data in real-time, and the historical
data set. It estimates the degradation trend, the reliability, and the life expectancy
(Zhao et al., 2021). The RUL analysis can be developed using statistical modeling and data-driven techniques. On the one hand, the data-driven methods aim
to learn and project the future to estimate the degradation. Support Vector Machine (SVM) (see Jia et al., 2020) and ANN (see Venkatesan et al., 2019) are
commonly used for this purpose. Still, it needs a large quantity of data, including the possible degradation trajectory of the component. Moreover, the training
and the learning occur offline, and it does not consider the changes over time,
which is unsuitable for a digital twin in the case of photovoltaics as it is under
dynamic environmental conditions (Zhong et al., 2023). On the other hand, statistical modeling can incorporate physical degradation and time variation into
the model. The main technique is the autoregression model, and its variations as
moving average (see Liu et al., 2023) and Gaussian regression (see Zhou et al.,
2018).

Laayouj et al. (2016) use a Relevance Vector Machine (RVM) for RUL in a
Photovoltaic system, as it has a high learning ability based on statistical probability learning. The challenges in RUL's prediction are the computation time
and the amount of data necessary to perform an acceptable RUL in a dynamic
environment. For instance, using Fuzzy evaluation to improve the initial computational efficiency together with Gaussian Process Regression which can lead to
high-dimension problems efficiently, can be a good solution for this type of application (Kang et al., 2020). Moreover, Liu et al. (2023) presented the semipara-
metric modeling of PV to determine the degradation based on dynamic covariate
data such as temperature, solar irradiance, and UV rays combining the degra-

dation model with multivariate Bernstein bases. However, the main challenge is forecasting the main parameters in the long term to provide an accurate RUL. Moreover, the ambient conditions' effect on the PV array and the converter is essential to identify and include in the RUL's prediction model and the operation modes. So far, there is no RUL analysis of the PV array together with the power converter as both depend on each other, and a holistic point of view could help to deploy an enhanced and more accurate RUL.

5.3. Decision-making

An important aspect of Digital Twin is the decision-making after the RUL and Fault diagnosis is developed. This is a new trend due to the IoT and Energy 4.0 development. Thus, novel decision-making can come in the coming years. Here, we will analyze the current industry decision-making paradigm and how it can be applied to PVDC.

Considering the real-time operation, its predictions, and future performance, DT can generate maintenance recommendations, manage the power converters accordingly, schedule PV surface cleaning, and replace capacitors and diodes before the failure happens. Also, it can help identify large-scale installations the fault PV panel. Thus, the decision-making process can be divided into three categories: (i) Maintenance Planning and Scheduling, (ii) Reliability and cost decision-making, (iii) Optimization management and control (Bousdekis et al., 2019).

- Maintenance Planning and Scheduling: this category includes data analysis of the current operation, its degradation, plus the fabric parameters. It can estimate the better dates to perform maintenance, like cleaning and replacing the converter's fans, capacitors, diodes, and breakers. Adequate maintenance planning helps to increase the life expectancy and efficiency of the PV system. Thanks to DT, this can be automatically done.
- Reliability and Cost Decision-making: In recent years, the semiconductors and chips supply chain suffered from extensive waiting time. Knowing the power converter's RUL in the DT paradigm could help buy the new device in advance without affecting the system's real-time operation. The same can happen with the photovoltaic panels, where the DT can inform the provider of a failure and its consequences on the degradation of the components and thus its replacement in a certain period of time. However, this should go hand in hand with the cost estimation
- Optimization Management and Control: due to degradation, some active power control can be triggered to reduce the damage to the PV system. Also, the energy management system could deactivate a PV string or a power converter in a large installation, as this could cause further failures.

The decision-making should take into consideration these categories but also uncertainties. Thus, multi-objective optimization can help with the decision problem. In industry, optimization is based on maximizing reliability under minimal operational cost. In this sense and considering uncertainties, multi-objective fuzzy optimization (see Kaplan and Can, 2021), two-stage optimization with multi-objective decision-making using genetic algorithms (see Li et al., 2009) could be a good approach for decision-making. Van Horenbeek and Pintelon (2013) proposed a dynamic schedule considering policies and the degradation of various components. However, in the field of PV, there is a lack of studies considering predictive maintenance, reliability, cost, and supply chain schedules (Peinado Gonzalo et al., 2020). With the integration of big data, IoT, and enhanced modeling techniques, the decision-making step in PVDC should improve in the coming years.

6. Challenges in implementing digital twins for PV microgrids

Digitalizing PVDC microgrids may offer various advantages, including preventive maintenance, enhanced circular economy, and increased efficiency. However, the path to arriving at a full digitalization may encounter various challenges regarding costs, data quality and availability, data privacy, accuracy and reliability, and connectivity. In this section, we analyze these challenges, focusing on paying the way for a digital PVDC microgrid.

- Operational Cost: In PV systems, the average operational cost for utility PV power plants in the US goes from 10 to 18 USD/kW/year, and in Europe, the cost is about 10 USD/kW. However, about 70% of the cost is due to preventive maintenance and cleaning (IRENA, 2020). This system's digitalization could help reduce the operational cost by 10 % (Clifton et al., 2023). However, implementing digital twins for PV can result in significant upfront costs and ongoing expenses that still need to be analyzed, which can cause a delay in digital twin adoption.
- Data Quality and Availability: Data quality and availability are fundamental to the success of a digital twin. Insufficient, inaccurate, or outdated data can lead to erroneous predictions and undermine the digital twin's effectiveness. Thus, it is necessary to develop techniques to achieve data quality by detecting incorrect data in real-time. For instance, Rodríguez et al. (2023) proposed a methodology to detect incorrect data from each of the sensors of their microgrid using artificial intelligence and analyzing it when there is an update between the physical and digital world. Their solution does not involve real-time analysis and integration between the physical components and the digital platform. Similar solutions are time-consuming and need high computational resources. Thus, the challenge is finding solutions to guarantee the data quality of the PVDC microgrid in real-time without compromising the speed or the computational resources.
- Data Privacy and Security: The use of digital twins in PV microgrids involves collecting and analyzing large volumes of sensitive data from the

generator and the user. So far, the protocols used for communication for this sensitive data are vulnerable to hacking, causing not only a leak of data but also the malfunction of the microgrid. It has been reported that cyber-attacks are more likely to target control systems to cause harm (Plėta et al., 2020). For this, it is necessary to address protocols that ensure quick communication and offer robust security, protecting data privacy and offering energy security.

- Accuracy and Trustworthiness: The accuracy and reliability of a digital twin depending on the calibration and validation against real-world performance and the digital counterpart. Thus, dynamic training and online autonomous calibration can be necessary. In PVDC microgrids, there is no reported study, but in the case of wind power plants, there are some calibration proposals to gain accuracy and trustworthiness (Song et al., 2022).

- Dependency on Connectivity: Digital twins heavily rely on a robust and reliable network infrastructure to facilitate real-time data transmission and remote control. In locations with limited connectivity or unreliable networks, ensuring uninterrupted communication can be a limiting factor. According to the European Union Agency for Cybersecurity (ENISA), Europe experienced 168 reported telecommunication incidents in 2021 that caused 5,106 million user lost hours due to natural disasters, malware and system failures, and human errors (Malatras et al., 2021).

Addressing these challenges is crucial for organizations considering the adoption of digital twins for PV microgrids. By conducting thorough cost-benefit analyses, implementing robust data privacy measures, and employing reliable data validation processes, PVDC microgrids can maximize the potential benefits while mitigating risks. Therefore, it is paramount to continuously evaluate the digital twin's performance and adaptability to ensure its sustainability and effectiveness.

7. Conclusions

This chapter discussed the PVDT for DC microgrids. It covered its framework, the PV panels' modeling techniques, the converter's role, and the supervisory control and management system. The following conclusions can be drawn:

- Due to the development of IoT, the integration between the Physical layer and the digital one can be possible. However, harmonizing the standards, the protocols, and the cybersecurity are currently a challenge to overcome in the communication layer. This can permit the response in an emergency, protect users' data, and enhance the reliability of the PVDC microgrid.

- This chapter presents various modeling techniques for PV panels and the DC-DC converter from physical to data-driven modeling that can be applied in real-time simulation and Digital twin framework. However, there is still a need for real applications using techniques that can adapt in the time for electrical modeling and thermal modeling. The two components degrade due to the operating condition, specifically the temperature. If the DT has the main objective of failure detection, then special attention should be given to thermal modeling.

- The converter is an integral part of the PVDC microgrid; without it, there cannot be an interaction between the PV panel, the load, and other energy generators. It helps to control the DC voltage and the active power and performs its internal control to operate the switches devices like IGBT or MOSFET. It can also help reduce the degradation of the various components by limiting the active power when quick solar irradiance variations occur or when the ambient temperature is high. The active power control not only focuses on the energy demand but also the DC bus and the power stability so the load or the other components of the DC microgrid do not suffer any damage due to solar irradiance variations.

- The integration of DT into the architecture of a SCADA enhances the supervision, monitoring, and management of the PVDC microgrid. From augmentative reality interface to health, management helps with decision-making and enhances control. It is necessary to improve the decision-making techniques for PVDC, as it can go from scheduled maintenance to a change of the power point of operation. However, fault detection and RUL lack a holistic analysis considering the PV panel and its converter. Any variation or fault in the PV panel could affect the operation and also the RUL in the converter. Thus, researchers and stakeholders should consider the two components as a unit to perform a better and enhanced real-time analysis.

- The chapter extensively explores the various challenges that PVDC microgrids may face as they progress toward future implementation. A crucial step in this process is conducting a comprehensive cost analysis to deeply understand their economic viability, thereby fostering greater stakeholder engagement and investment in DT for PVDC microgrids. Additionally, the chapter delves into the research and data policy challenges, addressing critical aspects such as data privacy, data quality, and the necessity for real-time calibration. Notably, the wind power sector has already addressed these challenges, whereas PVDC microgrids are still nascent, necessitating focused efforts to overcome these obstacles. Therefore, stakeholders, researchers, and policymakers should proactively tackle these challenges to help the integration and management of DT for PVDC microgrids.

Funding

This work was supported in part by the Ministero dell'Istruzione, dell'Università e della Ricerca (Italy) under Grant PRIN2020–HOTSPHOT 2020LB9TBC. The European Union also supported this work under the Italian National Recovery and Resilience Plan (NRRP) of NextGenerationEU, partnership on "Telecommunications of the Future" (PE00000001 - program "RESTART", Structural Project 6GWINET)). This manuscript reflects only the authors' views and opinions; neither the European Union nor the European Commission can be considered responsible for them.

References

Alexopoulos, K., Nikolakis, N. and Chryssolouris, G. 2020, 5. Digital twin-driven supervised machine learning for the development of artificial intelligence applications in manufacturing. https://doi.org/10.1080/0951192X.2020.1747642, 33(5): 429–439. Retrieved from https://www.tandfonline.com/doi/abs/10.1080/0 951192X.2020.1747642.

Ali, S., Zheng, Z., Aillerie, M., Sawicki, J. P., Péra, M. C. and Hissel, D. 2021, 7. A review of dc microgrid energy management systems dedicated to residential applications. Energies, 14(14).

Analytics, I. 2021. IoT 2021 in review: The 10 Most Relevant IoT Developments of the Year (Tech. Rep.). IoT analytics. Retrieved from https://iot-analytics.com/iot-2021-in-review/.

Andresen, M., Schloh, M., Buticchi, G. and Liserre, M. 2016. Computational light junction temperature estimator for active thermal control. ECCE 2016 - IEEE Energy Conversion Congress and Exposition, Proceedings.

Arafet, K. and Berlanga, R. 2021, 5. Digital twins in solar farms: An approach through time series and deep learning. Algorithms, 14(5).

Aslam, S., Herodotou, H., Mohsin, S. M., Javaid, N., Ashraf, N. and Aslam, S. 2021, 7. A survey on deep learning methods for power load and renewable energy forecasting in smart microgrids. Renewable and Sustainable Energy Reviews, 144.

Baba, A. O., Liu, G. and Chen, X. 2020, 1. Classification and evaluation review of maximum power point tracking methods. Sustainable Futures, 2: 100020.

Bartsch, K., Pettke, A., Höbert, A., Lakämper, J. and Lange, F. 2021, 7. On the digital twin application and the role of artificial intelligence in additive manufacturing: A systematic review. J. Phys. Materials, 4(3).

Bazmohammadi, N., Madary, A., Vasquez, J. C., Bazmohammadi, H., Khan, B., Wu, Y. and Guerrero, J. M. 2021. Microgrid digital twins: Concepts, applications, and future trends. IEEE Access.

Behera, M. K., Majumder, I. and Nayak, N. 2018, 6. Solar photovoltaic power forecasting using optimized modified extreme learning machine technique. Engineering Science and Technology, an International Journal, 21(3): 428–438.

Bendib, B., Krim, F., Belmili, H., Almi, M. F. and Boulouma, S. 2014, 1. Advanced fuzzy MPPT controller for a stand-alone PV system. Energy Procedia, 50: 383–392.

Beriber, D. and Talha, A. 2013, 5. MPPT techniques for PV systems. In 4th International Conference on Power Engineering, Energy and Electrical Drives, pp. 1437–1442. IEEE. Retrieved from http://ieeexplore.ieee.org/articleDetails. jsp?arnumber=6635826.

Bindi, M., Corti, F., Aizenberg, I., Grasso, F., Lozito, G. M., Luchetta, A., . . . Reatti, A. 2022, 3. Machine learning-based monitoring of dc-dc converters in photovoltaic applications. Algorithms, 15(3).

Borni, A., Abdelkrim, T., Bouarroudj, N., Bouchakour, A., Zaghba, L., Lakhdari, A. and Zarour, L. 2017, 7. Optimized MPPT controllers using GA for grid connected photovoltaic systems, comparative study. Energy Procedia, 119: 278–296.

Botín-Sanabria, D. M., Mihaita, S., Peimbert-García, R. E., Ramírez-Moreno, M. A., Ramírez-Mendoza, R. A. and Lozoya-Santos, J. D. J. 2022, 3. Digital twin technology challenges and applications: A comprehensive review. Remote Sensing, 14(6).

Bousdekis, A., Lepenioti, K., Apostolou, D. and Mentzas, G. 2019, 1. Decision making in predictive maintenance: Literature review and research agenda for Industry 4.0. IFACPapersOnLine, 52(13): 607–612.

Cabrera-Tobar, A., Blasuttigh, N., Massi Pavan, A., Lughi, V., Petrone, G. and Spagnuolo, G. 2022, 11. Energy scheduling and performance evaluation of an e-vehicle charging station. Electronics, 11(23): 3948.

Cabrera-Tobar, A., Bullich-Massagué, E., Aragüés-Peñalba, M. and Gomis-Bellmunt, O. 2016, 6. Topologies for large scale photovoltaic power plants. Renewable and Sustainable Energy Reviews, 59: 309–319. Retrieved from http:// www.sciencedirect.com/science/article/pii/S1364032116000289.

Cabrera-Tobar, A., Bullich-Massagué, E., Aragüés-Peñalba, M. and Gomis-Bellmunt, O. 2019, 10. Active and reactive power control of a PV generator for grid code compliance. Energies, 12(20): 3872. Retrieved from https://www.mdpi. com/1996-1073/12/20/3872.

Cabrera-Tobar, A., Gomis-Bellmunt, O., Carrasco, J. and Barnes, M. 2018, 6. Evaluation of a state observer for frequency estimation in a grid tied photovoltaic inverter. In 2018 IEEE International Conference on Environment and Electrical Engineering and 2018 IEEE Industrial and Commercial Power Systems Europe (EEEIC/I&CPS Europe), pp. 1–6. IEEE. Retrieved from https://ieeexplore.ieee. org/document/8494620/.

Cabrera Tobar, A. K. 2018. Large scale photovoltaic power plants: Configuration, integration and control (Doctoral dissertation). Retrieved from https://upcommons. upc.edu/handle/2117/120996.

Cai, Y., Lin, P., Lin, Y., Zheng, Q., Cheng, S., Chen, Z. and Wu, L. 2020, 2. Online photovoltaic fault detection method based on data stream clustering. IOP Conference Series: Earth and Environmental Science, 431(1): 012060. Retrieved from https://iopscience.iop.org/article/10.1088/1755-1315/431/1/012060; https://iopscience.iop.org/article/10.1088/1755-1315/431/1/012060/meta.

Cavallaro, D., Pulvirenti, M., Zanetti, E. and Saggio, M. 2019. Capability of SiC MOSFETs under short-circuit tests and development of a thermal model by finite element analysis. Materials Science Forum, 963: 788–791. Retrieved from https://www.scientific.net/MSF.963.788.

Chine, W., Mellit, A., Lughi, V., Malek, A., Sulligoi, G. and Massi Pavan, A. 2016, 5. A novel fault diagnosis technique for photovoltaic systems based on artificial neural networks. Renewable Energy, 90: 501–512.

Clifton, A., Barber, S., Bray, A., Enevoldsen, P., Fields, J., Sempreviva, A. M., . . . Ding, Y. 2023, 6. Grand challenges in the digitalisation of wind energy. Wind Energy Science, 8(6): 947–974.

Dolara, A., Leva, S. and Manzolini, G. 2015, 9. Comparison of different physical models for PV power output prediction. Solar Energy, 119: 83–99.

Femia, N., Petrone, G., Spagnuolo, G. and Vitelli, M. 2012. Power Electronics and Control Techniques for Maximum Energy Harvesting in Photovoltaic Systems. CRC Press. Retrieved from https://www.taylorfrancis.com/books/mono/10.1201/b14303/power-electronics-control-techniques-maximum.

Frances, A., Asensi, R., Garcia, O., Prieto, R. and Uceda, J. 2018. Modeling electronic power converters in smart dc microgrids—An overview. IEEE Transactions on Smart Grid, 9(6): 6274–6287.

Gao, Y., Xu, G., Wang, S., Li, Y., Cai, R., Li, Z., . . . Zhang, Y. 2022. Solar array power prediction for near-space vehicles based on a hybrid method. IEEE Access, 10: 123233–123250.

Glaessgen, E. H. and Stargel, D. S. 2012. The Digital Twin Paradigm for Future NASA and U.S. Air Force Vehicles.

Hacı Bektaş, N., Üniversitesi, V., Fakültesi, M. -M. and Bölümü, E.-E. M. 2021, 9. Digital twin concept for renewable energy sources. Konya Journal of Engineering Sciences, 9(3): 836–844. Retrieved from https://dergipark.org.tr/en/pub/konjes/issue/64647/969989.

Haider, S. I., Burton, L. and Joshi, Y. 2010, 2. A proper orthogonal decomposition based system-level thermal modeling methodology for shipboard power electronics cabinets. https://doi.org/10.1080/01457630701686743, 29(2): 198–215. Retrieved from https://www.tandfonline.com/doi/abs/10.1080/01457630701686743.

Hasan, M. K., Habib, A. K., Islam, S., Safie, N., Abdullah, S. N. H. S. and Pandey, B. 2023, 10. DDoS: Distributed denial of service attack in communication standard vulnerabilities in smart grid applications and cyber security with recent developments. Energy Reports, 9: 1318–1326.

Hu, G., Cai, T., Chen, C. and Duan, S. 2009, 5. Solutions for SCADA system communication reliability in photovoltaic power plants. In 2009 IEEE 6th

International Power Electronics and Motion Control Conference, pp. 2482–2485. IEEE. Retrieved from http://ieeexplore.ieee.org/lpdocs/epic03/wrapper.htm?arnumber=5157821.

IRENA. 2020. Renewable Power Generation Costs in 2020 (Tech. Rep.). Retrieved from https://www.irena.org/.

Jain, P., Poon, J., Singh, J. P., Spanos, C., Sanders, S. R. and Panda, S. K. 2020, 1. A digital twin approach for fault diagnosis in distributed photovoltaic systems. IEEE Transactions on Power Electronics, 35(1): 940–956.

Jia, J., Liang, J., Shi, Y., Wen, J., Pang, X. and Zeng, J. 2020. SOH and RUL prediction of Lithium-Ion batteries. Energies, 13: 375.

Kang, W., Xiao, J., Xiao, M., Hu, Y., Zhu, H. and Li, J. 2020. Research on remaining useful life prognostics based on fuzzy evaluation-gaussian process regression method. IEEE Access, 8: 71965–71973.

Kaplan, H. and Can, M. 2021, 2. Fuzzy reliability theory in the decision-making process. Advancements in Fuzzy Reliability Theory, 76–89.

Kondoro, A., Ben Dhaou, I., Tenhunen, H. and Mvungi, N. 2021, 3. Real time performance analysis of secure IoT protocols for microgrid communication. Future Generation Computer Systems, 116: 1–12.

Kritzinger, W., Karner, M., Traar, G., Henjes, J. and Sihn, W. 2018, 1. Digital Twin in manufacturing: A categorical literature review and classification, 51(11): 1016–1022.

Kumar, A. and Singh, A. 2017. Stream mining a review: Tool and techniques. Proceedings of the International Conference on Electronics, Communication and Aerospace Technology, ICECA, 27–32.

Laayouj, N., Jamouli, H. and El Hail, M. 2016, 11. Photovoltaic module health monitoring and degradation assessment. Conference on Control and Fault-Tolerant Systems, SysTol, 2016-Novem: 585–592.

Leva, S., Mussetta, M. and Ogliari, E. 2019, 5. PV module fault diagnosis based on microconverters and day-ahead forecast. IEEE Transactions on Industrial Electronics, 66(5): 3928–3937.

Li, Z., Liao, H. and Coit, D. W. 2009, 10. A two-stage approach for multi-objective decision making with applications to system reliability optimization. Reliability Engineering & System Safety, 94(10): 1585–1592.

Liu, Q., Hu, Q., Zhou, J., Yu, D. and Mo, H. 2023. Remaining useful life prediction of PV systems under dynamic environmental conditions. IEEE Journal of Photovoltaics, 1–13. Retrieved from https://ieeexplore.ieee.org/document/10123306/.

Lu, J., Liu, A., Dong, F., Gu, F., Gama, J. and Zhang, G. 2019, 12. Learning under concept drift: A review. IEEE Transactions on Knowledge and Data Engineering, 31(12): 2346–2363.

Malatras, A., Bafoutsou, G., Taurins, E., Dekker, M. and Cybersecurity, E. U. A. f. 2021. ENISA - Annual report telecom security incidents 2021 (Tech. Rep. No. July). ENISA.

Manowska, A., Wycisk, A., Nowrot, A. and Pielot, J. 2022, 12. The use of the MQTT protocol in measurement, monitoring and control systems as part of the implementation of energy management systems. Electronics, 12(1): 17. Retrieved from https://www.mdpi.com/2079-9292/12/1/17/htm; https://www.mdpi.com/2079-9292/12/1/17.

Markets and Markets. 2020. Electrical Digital Twin Market Growth Drivers & Opportunities (Tech. Rep.). Market and Markets. Retrieved from https://www.marketsandmarkets.com/Market-Reports.

Markvart, T. 2012. Practical Handbook of Photovoltaics. Elsevier. Retrieved from http://www.sciencedirect.com/science/article/pii/B9780123859341000209.

Massidda, L. and Marrocu, M. 2017, 4. Use of multilinear adaptive regression splines and numerical weather prediction to forecast the power output of a PV plant in Borkum, Germany. Solar Energy, 146: 141–149.

Mellit, A., Pavan, A. M., Ogliari, E., Leva, S. and Lughi, V. 2020. Advanced methods for photovoltaic output power forecasting: A review. Applied Sciences (Switzerland), 10(2): 1–23.

Migliorini, L., Molinaroli, L., Simonetti, R. and Manzolini, G. 2017, 3. Development and experimental validation of a comprehensive thermoelectric dynamic model of photovoltaic modules. Solar Energy, 144: 489–501.

Mihaylov, B., Betts, T. R., Pozza, A., Mullejans, H. and Gottschalg, R. 2016, 11. Uncertainty Estimation of temperature coefficient measurements of PV modules. IEEE Journal of Photovoltaics, 6(6): 1554–1563.

Milton, M., De La, C. O., Ginn, H. L. and Benigni, A. 2020, 9. Controller-embeddable probabilistic real-time digital twins for power electronic converter diagnostics. IEEE Transactions on Power Electronics, 35(9): 9852–9866.

Modu, B., Abdullah, M. P., Sanusi, M. A. and Hamza, M. F. 2023, 5. DC-based microgrid: Topologies, control schemes, and implementations. Alexandria Engineering Journal, 70: 61–92.

Mosaad, M. I., Osama abed el Raouf, M., Al-Ahmar, M. A. and Banakher, F. A. 2019, 4. Maximum power point tracking of PV system based Cuckoo search algorithm; review and comparison. Energy Procedia, 162: 117–126.

Mustafa, M. A., Cleemput, S., Aly, A. and Abidin, A. 2019, 11. A secure and privacy-preserving protocol for smart metering operational data collection. IEEE Transactions on Smart Grid, 10(6): 6481–6490.

Nahvi, S. A., Bazaz, M. A. and Khan, H. 2017, 12. Model order reduction in power electronics: Issues and perspectives. Proceeding—IEEE International Conference on Computing, Communication and Automation, ICCCA 2017, 2017-Janua: 1417–1421.

Nguimfack-Ndongmo, J. D. D., Ngoussandou, B. P., Goron, D., Ajesam Asoh, D., Kidmo, D. K., Mbaka Nfah, E. and Kenné, G. 2022, 11. Nonlinear neuro-adaptive MPPT controller and voltage stabilization of PV Systems under real environmental conditions. Energy Reports, 8: 1037–1052.

Niccolai, A., Dolara, A. and Grimaccia, F. 2018, 12. Analysis of photovoltaic five-parameter model. Proceedings of International Conference on Smart Systems and Technologies 2018, SST 2018: 205–210.

Pargmann, H., Euhausen, D. and Faber, R. 2018, 6. Intelligent big data processing for wind farm monitoring and analysis based on cloud-Technologies and digital twins: A quantitative approach. 2018 3rd IEEE International Conference on Cloud Computing and Big Data Analysis, ICCCBDA, 2018: 233–237.

Peinado Gonzalo, A., Pliego Marugán, A. and García Márquez, F. P. 2020, 12. Survey of maintenance management for photovoltaic power systems. Renewable and Sustainable Energy Reviews, 134: 110347.

Piazza, M. C. D., Luna, M., Petrone, G. and Spagnuolo, G. 2017, 7. Translation of the single-diode PV model parameters identified by using explicit formulas. IEEE Journal of Photovoltaics, 7(4): 1009–1016.

Piliougine, M., Guejia-Burbano, R. A. and Spagnuolo, G. 2022. Detecting partial shadowing and mismatching phenomena in photovoltaic arrays by machine learning techniques. IEEE Open Journal of the Industrial Electronics Society, 3: 507–521.

Piliougine, M., Oukaja, A., Sidrach-de Cardona, M. and Spagnuolo, G. 2021, 5. Temperature coefficients of degraded crystalline silicon photovoltaic modules at outdoor conditions. Progress in Photovoltaics: Research and Applications, 29(5): 558–570.

Planas, E., Andreu, J., Gárate, J. I., Martínez De Alegría, I. and Ibarra, E. 2015. AC and DC technology in microgrids: A review. Renewable and Sustainable Energy Reviews, 43: 726–749.

Plėta, T., Tvaronavičienė, M., Della Casa, S. and Agafonov, K. 2020, 9. Cyber-attacks to critical energy infrastructure and management issues: Overview of selected cases. Insights into regional development. Vilnius: Entrepreneurship and Sustainability Center, 2(3): 703–715. Retrieved from http://jssidoi.org/IRD/http://doi.org/10.9770/.

Preprint, E., Fu, Y., Gao, Z. and Zhang, A. 2021. Data-Driven Parameter Fault Classification for A DC-DC Buck Converter (Tech. Rep.).

Qais, M. H., Hasanien, H. M., Alghuwainem, S., Loo, K. H., Elgendy, M. A. and Turky, R. A. 2022, 5. Accurate Three-Diode model estimation of Photovoltaic modules using a novel circle search algorithm. Ain Shams Engineering Journal, 13(3).

Rodríguez, F., Chicaiza, W. D., Sánchez, A. and Escaño, J. M. 2023, 10. Updating digital twins: Methodology for data accuracy quality control using machine learning techniques. Computers in Industry, 151: 103958.

Rouzbehi, K., Miranian, A., Escaño, J. M., Rakhshani, E., Shariati, N. and Pouresmaeil, E. 2019, 5. A data-driven based voltage control strategy for DC-DC converters: Application to DC microgrid. Electronics (Switzerland), 8(5).

Sadabadi, M. S., Sahoo, S. and Blaabjerg, F. 2022, 2. Stability-oriented design of cyberattack-resilient controllers for cooperative DC microgrids. IEEE Transactions on Power Electronics, 37(2): 1310–1321.

Samanta, H., Bhattacharjee, A., Pramanik, M., Das, A., Bhattacharya, K. D. and Saha, H. 2020, 12. Internet of things based smart energy management in a vanadium redox flow battery storage integrated bio-solar microgrid. Journal of Energy Storage, 32: 101967.

Santos, L. D. O., De Carvalho, P. C. M. and Filho, C. D. O. C. 2022, 1. Photovoltaic cell operating temperature models: A review of correlations and parameters. IEEE Journal of Photovoltaics, 12(1): 179–190.

Sarmas, E., Dimitropoulos, N., Marinakis, V., Mylona, Z. and Doukas, H. 2022, 8. Transfer learning strategies for solar power forecasting under data scarcity. Scientific Reports, 12(1): 1–13. Retrieved from https://www.nature.com/articles/s41598-022-18516-x.

Schreiber, J. and Sick, B. 2022, 4. Model selection, adaptation, and combination for transfer learning in wind and photovoltaic power forecasts. Energy and AI, 100249. Retrieved from https://arxiv.org/abs/2204.13293v3.

Siddiqui, M. U. and Arif, A. F. 2013. Electrical, thermal and structural performance of a cooled PV module: Transient analysis using a multiphysics model. Applied Energy, 112: 300–312.

Song, H., Song, M. and Liu, X. 2022, 11. Online autonomous calibration of digital twins using machine learning with application to nuclear power plants. Applied Energy, 326: 119995.

Soubdhan, T., Ndong, J., Ould-Baba, H. and Do, M. T. 2016, 6. A robust forecasting framework based on the Kalman filtering approach with a twofold parameter tuning procedure: Application to solar and photovoltaic prediction. Solar Energy, 131: 246–259.

Stella, F., Pellegrino, G., Armando, E. and Dapra, D. 2018, 7. Online junction temperature estimation of Sic power MOSFETS through on-state voltage mapping. IEEE Transactions on Industry Applications, 54(4): 3453–3462.

Sukumar, S., Marsadek, M., Agileswari, K. R. and Mokhlis, H. 2018, 12. Ramp-rate control smoothing methods to control output power fluctuations from solar photovoltaic (PV) sources—A review. Journal of Energy Storage, 20: 218–229.

Tanjimuddin, M., Kannisto, P., Jafary, P., Filppula, M., Repo, S. and Hästbacka, D. 2022, 12. A comparative study on multi-agent and service-oriented microgrid automation systems from energy internet perspective. Sustainable Energy, Grids and Networks, 32: 100856.

Tanyingyong, V., Olsson, R., Cho, J. W., Hidell, M. and Sjodin, P. 2016. IoT-Grid: IoT communication for smart DC grids. 2016 IEEE Global Communications Conference, GLOBECOM 2016 - Proceedings.

Tao, F., Zhang, H., Liu, A. and Nee, A. Y. 2019, 4. Digital twin in industry: State-of-the-Art. IEEE Transactions on Industrial Informatics, 15(4): 2405–2415.

Van Horenbeek, A. and Pintelon, L. 2013, 12. A dynamic predictive maintenance policy for complex multi-component systems. Reliability Engineering & System Safety, 120: 39–50.

Venkatesan, S., Manickavasagam, K., Tengenkai, N. and Vijayalakshmi, N. 2019. Health monitoring and prognosis of electric vehicle motor using intelligent-digital twin. IET Electric Power Applications, 13(9): 1328–1335.

Wunderlich, A. and Santi, E. 2021, 6. Digital twin models of power electronic converters using dynamic neural networks. Conference Proceedings PEC, 2369–2376.

Yang, Y., Yu, T., Zhao, W. and Zhu, X. 2021, 9. Kalman Filter photovoltaic power prediction model based on forecasting experience. Frontiers in Energy Research, 9: 472.

Yap, K. Y., Sarimuthu, C. R. and Lim, J. M. Y. 2020, 11. Artificial intelligence based MPPT techniques for solar power system: A review. Journal of Modern Power Systems and Clean Energy, 8(6): 1043–1059.

Yu, Y., Liu, G. P. and Hu, W. 2022, 5. Blockchain protocol-based secondary predictive secure control for voltage restoration and current sharing of DC microgrids. IEEE Transactions on Smart Grid.

Yuan, G. and Xie, F. 2023, 1. Digital Twin-Based economic assessment of solar energy in smart microgrids using reinforcement learning technique. Solar Energy, 250: 398–408.

Zanatta, N., Bignucolo, F. and Cabrera-Tobar, A. 2017. Active power control of a PV generator for large scale PV power plants (Unpublished doctoral dissertation). University of Padova.

Zhang, S., Tan, W. and Li, Y. 2018, 6. A survey of online sequential extreme learning machine. 2018 5th International Conference on Control, Decision and Information Technologies, CoDIT, 2018: 45–50.

Zhang, T., Zhang, W., Zhao, Q., Chen, J. and Zhang, Y. 2023. Adaptive particle filter with randomized quasi-monte carlo sampling for unbalanced distribution system state estimation. IEEE Transactions on Instrumentation and Measurement, 1–1. Retrieved from https://ieeexplore.ieee.org/document/10124387/.

Zhang, Y., Wang, Z., Wang, H. and Blaabjerg, F. 2020, 10. Artificial intelligence-aided thermal model considering cross-coupling effects. IEEE Transactions on Power Electronics, 35(10): 9998–10002.

Zhao, S., Blaabjerg, F. and Wang, H. 2021, 4. An overview of artificial intelligence applications for power electronics. IEEE Transactions on Power Electronics, 36(4): 4633–4658.

Zhong, D., Xia, Z., Zhu, Y. and Duan, J. 2023, 4. Overview of predictive maintenance based on digital twin technology. Heliyon, e14534.

Zhou, D., Yin, H., Fu, P., Song, X., Lu, W., Yuan, L. and Fu, Z. 2018. Prognostics for state of health of lithium-ion batteries based on Gaussian process regression. Mathematical Problems in Engineering, 2018.

Zia, M. F., Benbouzid, M., Elbouchikhi, E., Muyeen, S. M., Techato, K. and Guerrero, J. M. 2020. Microgrid transactive energy: Review, architectures, distributed ledger technologies, and market analysis. IEEE Access, 8: 19410–19432.

CHAPTER 5

From Microgrids to Virtual Power Plants: A Cybersecurity Perspective

Giovanni Gaggero, Diego Piserà, Paola Girdinio, Federico Silvestro*
and *Mario Marchese*

1. Introduction

Over the past twenty years, advancements in renewable energy sources such as distributed generation units, as well as the implementation of energy storage systems, flexible AC transmission systems, active demand management, AC microgrids, and innovative control strategies relying on information and communication technologies (ICTs), have enabled energy professionals and researchers to reimagine traditional power systems. Microgrids represented one of the key technology in this change. Despite some partially-solved technical issues, with more distributed generation units that generate DC power, DC-based microgrid systems could soon be the right candidates for the future energy systems (Justo et al. (2013)). Microgrids offer an excellent solution for areas in the world where the primary grid remains underdeveloped. However, if a user has access to a robust distribution grid, they will likely operate primarily in a grid-connected mode, but can still utilize the islanded mode in the event of main grid issues to enhance resilience, maintain power continuity, or gain economic benefits. Constructing suitable infrastructure is a crucial practical challenge in harnessing the potential of DERs and controllable loads. Consider a scenario where a company possesses numerous properties or plots of land, each containing various generators and

Department of Electrical, Electronic and Telecommunications Engineering, and Naval Architecture (DITEN), University of Genoa, Via all'Opera Pia 11A, 16145, Genoa, Italy.
* Corresponding author: giovanni.gaggero@unige.it

loads. Connecting these areas with dedicated cables may prove impractical. As a result, power systems are moving towards coordinating distributed generation and loads utilizing public infrastructure for both communication and power supply purposes. One of the main examples of these technologies is represented by Virtual Power Plants (VPP). Virtual Power Plants are a concept in the energy industry that refers to a network of decentralized power generating units that are interconnected and managed as a unified system. Rather than relying on a single large power plant, VPPs harness the collective capacity of various smaller, distributed energy resources to function as a flexible and dynamic power generation and distribution system.

The main contribution of this paper is to analyze the evolution of the technologies employed to control power distribution systems from a cybersecurity perspective. This includes the evaluation of the attack surface with respect to novel network architectures, and the analysis of the potential impact of attacks on these systems. Also, this work provides a preliminary framework to deploy proper cybersecurity monitoring of these infrastructures. This work is structured as follows. Section 2 analyzes the state of the art regarding cybersecurity issues in the smart grid, with a particular focus on power distribution systems. Section 3 discusses the evolution of distribution power systems from microgrids to recent technologies such as Renewable Energy Communities (REC) and Virtual Power Plants (VPP). Section 4 analyzes the network architectures that are employed to properly control these systems, evaluating the whole attack surface and the risks associated with attacks on these infrastructures. Section 5 proposes a preliminary framework to implement a proper cybersecurity monitoring for these infrastructures that should be included into the enterprise event management of energy utilities. Finally, in Section 6 conclusions are drawn.

2. Related works

The issue of cybersecurity is becoming increasingly important in the power sector. This study focuses on a specific area within the smart grid environment. Several works in literature have taken into account the problem of cybersecurity in the smart grid environment, such as in Wang and Lu (2013) and in El Mrabet et al. (2018). The use of information and communication technologies is becoming increasingly prevalent in the rapidly evolving distribution grid environment. Critical components of the smart distribution grid are the Distributed Energy Resources (DER). The use of DER is on the rise, with an increase in the number and variety of stakeholders and the complexity of interactions between them. This complexity leads to serious security threats to the power grid. A hierarchically structured model is proposed in (NESCOR), based on five levels:

- Level 1 Autonomous DER Generation and Storage: It includes DER systems interconnected to the utility grid that usually operate autonomously according to pre-established settings.

- Level 2 Facilities DER Energy Management: A facility DER management system manages the operation of the Level 1 DER systems, modifying the settings of the autonomous DERs in order to coordinate them at a local level.
- Level 3 Utility and Retail Energy Provider Operational Communications: Extends beyond the local site to allow utilities to request or require DER systems to take specific actions.
- Level 4 Distribution Utility Operational Analysis: Is represented by utility applications that monitor the power system and assess if efficiency, reliability, or market advantage can be improved by having DER systems modify their operation.
- Level 5 Transmission and Market Operations: It involves Transmission System Operators, which may need to exchange information about the capabilities and operational status of larger DER systems and/or aggregated DER systems.

This model can be generically applied to all the technologies described in the next sections but does not specifically takes into account the most recent advancements in power distribution grids. Smart inverters are among the most critical components for the optimal functioning of different technologies such as Energy Communities. Tuyen et al. (2022) present a comprehensive review of the system structure and vulnerabilities of a typical inverter-based power system integrated with distributed energy resources. cyb propose a framework to safeguard DER from cyberattacks to maintain the stability and reliability of the grid.

Cybersecurity monitoring for the smart grid is a lively field of research. Radoglou-Grammatikis and Sarigiannidis (2019) present a comprehensive survey of Intrusion Detection and Prevention Systems designed to protect the Smart Grid environment. Still, even well-established solutions have to be adapted to novel technologies, and state-of-the-art lacks of solutions specifically designed for some of these technologies such as energy communities or virtual power plants. Security Information and Event Management systems have been widely deployed in several sectors, including the smart grid (Radoglou-Grammatikis et al. (2021)), since they are powerful tools to prevent and detect cyber-attacks (González-Granadillo et al. (2021)), but also to react, thanks to the integration of the SIEM with others tools such as the Security Orchestration Automation and Response (SOAR). The integration of Operational Technology-related data to obtain complete visibility of the industrial plants is still an ope issue; in particular, it is not clear how to integrate the data coming from the technologies discussed in the next sections. To the best of our knowledge, this is the first work addressing the issue of cybersecurity monitoring for novel technologies in smart power distribution systems.

3. From microgrids to virtual power plant

With the term Microgrid, we usually refer to a local electrical grid with defined electrical boundaries, acting as a single and controllable entity. Microgrids can be grid-connected, or isolated from the main grid. In the first case, the microgrid can decide to operate both in grid-connected mode, so that the main grid provides the frequency and voltage reference, or islanded, to reach economic or technical goals. In the second case, the microgrid has the only option to work as an isolated entity, providing all the necessary regulation of the grid. Isolated microgrids represent a great technology for those areas of the world in which the main grid has not been yet properly developed. On the contrary, if an user disposes of a connection to a strong distribution grid, it will probably operate mainly as grid-connected for most of the time, but can still make use of the islanded mode in case of problems on the main grid, improving the resilience and the power continuity, or to have an economic advantage. An economic advantage can be pursued through a smart coordination of distributed energy resources and controllable loads. For this reason, Energy Management Systems (EMS) are an extremely important field of research. EMS can collect different types of data, including weather forecasting, the current price of energy, habits of the users and so on, in order to make predictions and send commands to dispatchable generators and controllable loads. The variety of application fields is huge: the main devices include Electrical Storage Systems, Electric Vehicles, Heat Ventilation and Air Conditioning (HVAC) systems, but also household appliances.

One of the main practical issues for exploiting the potentialities of DERs and controllable loads is to build proper infrastructure. For example, it is possible that a company owns multiple buildings or land, in which it spreads different generators and loads, but it is practically unfeasible to put dedicated cables between these areas. Another example is vehicle charging stations: a company may own charging stations over a huge area, but cannot build a dedicated communication infrastructure between these devices and the EMS. Still, if these devices can coordinate themselves, this would produce advantages both for the end users and the distribution and transmission operators. The trend in power systems is therefore to coordinate distributed generation and loads by using the public infrastructure, both from the power and communication perspective.

From power perspective a first step in this direction was done through Active Distribution Systems (ADNs), which have evolved from traditional electric distribution networks. ADNs are implemented to address technical challenges arising from Distributed Energy Resources (DERs), such as uncertainty related to changes in distributed renewable generation. The variability of DERs can result in sudden and unpredictable effects on load variation, disconnections, and increased risks of overcurrent, short circuit, and out-of-range voltage. ADNs are considered active as they employ systems, such as Supervisory Control and Data Acquisition (SCADA) and Distribution Management Systems, to observe and control DERs and prevent such technical issues. Distribution network automation includes control center information systems, substation automation, and customer interfaces like Automated Meter Reading (AMR) energy meters. ADN operations involve collecting and utilizing a wide range of information from various stakeholders in the distribution system, including transmission system operators, energy retailers, customers, and local communities. A second step was the introduction of a new market actor, the aggregator, in the energy environment. Aggregators are responsible for aggregating distributed energy resources connected in a particular area of the electrical grid and coordinating them to maximize economic returns, for example by providing balancing services to the transmission grid or selling energy during peak demand periods. To coordinate DERs, aggregators use software tools called Virtual Power Plants, which allow them to receive real-time data from field systems and send control signals to energy resources. Similar to microgrids, virtual power plants use EMS to calculate the optimal dispatch of DERs. The EMS collects information from third-party sources via the internet, such as weather forecasts or energy market prices, and uses the internet to communicate with electricity market operators.

We are witnessing a major paradigm shift where the controls and logics previously used in microgrids are now being applied to DERs connected to the public distribution grid through VPPs, creating a system of devices that collaborate with each other through virtual connections.

An overview of this change of paradigm is shown in Figure 2. In the first case, the boundaries of the systems are clearly defined: the power system has one or multiple Points Of Delivery (POD), usually equipped with a smart meter, that allow the exchange of energy between the microgrid and the DSO; also from a communication point of view, a microgrid utilizes a SCADA systems, usually based on wired communication, whose architecture, following the well-known models such as the Purdue Model, identify a single communication with the external internet, being protected by a firewall. In the second case, the devices exchange energy through the public grid after establishing an agreement with the DSO, and communicate by using the public internet; in this case, the EMS can run on cloud platforms managed by third-part entities.

Two common examples of commercial technologies that follow this new paradigm are Commercial Virtual Power Plants and digital platforms for energy communities.

3.1. Renewable energy communities

RECs derives from the Renewable Energy Directive (2018/2001/EU) of the European Union, that proposes a legal framework for the development of renewable energy sources and citizen participation in the energy transition through two new instruments: collective self-consumption schemes (CSCs) and renewable energy communities (RECs). In practice, an Energy Community is an association that produces and shares renewable energy autonomously by using the public distribution grid, that can be composed of local citizens, businesses, public adminis-

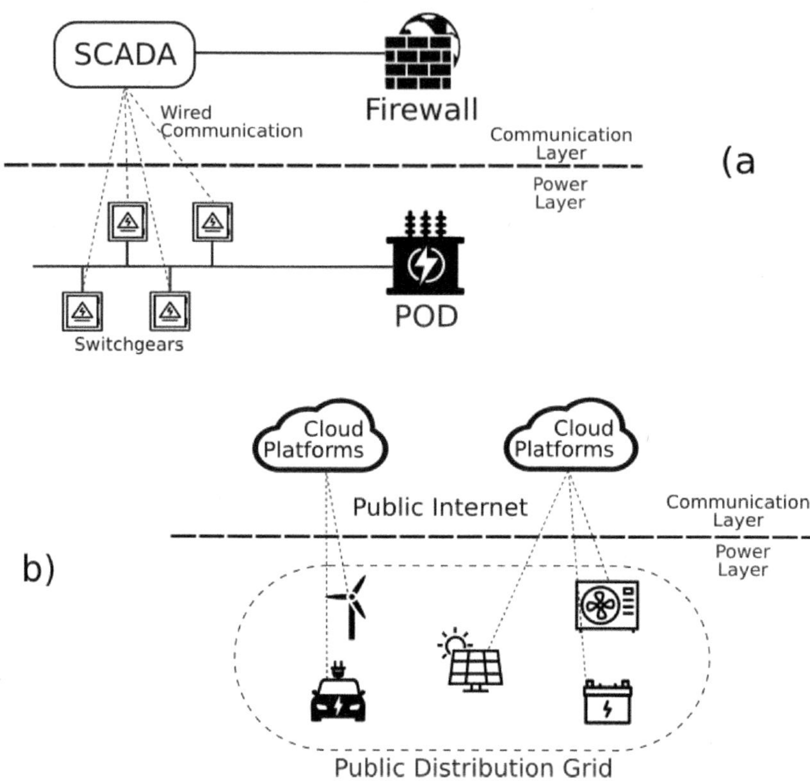

Figure 1. The evolution of power and network architectures from microgrids to smart power distribution systems.

trations, small and medium-sized enterprises, etc. Basically, any public or private entity can become a member. Once the generators are up and running, the Community can request to the DSO the incentives established by law for the shared energy, that is defined as energy consumed by members at the same time the energy is produced. In order to function properly, RECs make use of dedicated software platforms, that collect data, such as real-time electricity consumption, and control elements such as heat pumps and battery energy storage systems. The role of these platforms include dashboards for administration, real-time monitoring of REC and members, including the computation of treasury, and control though and EMS. Pereira et al. (2021) propose a Web-based platform for the management of energy communities for managing end-users' coordination, demand-response and energy tarifs. Chudoba and Borges (2022) discuss the role of scalable software solutions to maximize the value of distributed green assets, including distributed generation and HVAC. Moreover, we can already can find several commercial solutions for such platforms, such as in ROS and in HIV As discussed, it is not possible to build a dedicated infrastructure for the communication as well. The communication between the platform and the user must be done through the public internet. In this sense, we can say that the system EMS

- members of the REC is, from a control point of view, a SCADA system built on the public network.

3.2. Commercial virtual power plants

Commercial Virtual Power Plants (CVPPs) are digital platforms used by aggregators to manage a portfolio of medium and small Distributed Energy Resources (DERs), connected to the same portion of the grid; CVPPs allow DERs to participate in energy and balancing services markets in an aggregated form as if they were a single large power plant connected to the transmission grid. The aggregation of multiple DERs allows the minimum capacity thresholds required to participate in the wholesale energy market to be reached, and the diverse nature of the aggregated DERs reduces the risk of imbalance associated with a single resource (e.g., inaccurate forecasting for photovoltaic and wind turbine plants or failures of controllable resources such as storage or microturbines). CVPPs have an Energy Management System (EMS) that calculates the optimal energy quantities and prices to offer in energy and/or balancing markets, and the generation and/or absorption schedules that the controllable DERs must follow (Pudjianto et al. (2007)).

The EMS takes as input data coming from different sources, including: the marginal cost and operational parameters of individual DERs, internal forecasting algorithms used to forecast market prices and non-dispatchable renewable generation, data received from both field devices equipped with DERs (i.e., smart meters), third-party web services (such as market operator transparency platforms where auction results and prices are published), weather forecasting data shops (where weather data is used to forecast both non-dispatchable generation and price forecasting) and platforms to participate in market sessions (where the CVPP submits offers) (Rouzbahani et al. (2021)).

In the same portion of the grid, there may be DERs operated by different aggregators and therefore by different CVPPs, but the responsibility for operating the grid without technical problems remains with the DSO. The DSO, in turn, can use a VPP, called Technical Virtual Power Plants (TVPPs), to verify that the dispatching programs of the CVPPs do not cause technical problems to the electrical grid. TVPPs require information such as network topology and network flow history to perform the necessary checks, so they are operated by DSOs and not by aggregators. In the event that the DSO uses a TVPP, the aggregators operating in the portion of the grid must provide a communication channel between their CVPP and the DSO's TVPP (Pourghaderi et al. (2018)).

A real-life example of this technology is Tesla's VPP, used in the summer of 2022 in California to address grid instabilities created by heatwaves. Tesla used a digital platform to aggregate and coordinate Powerwall batteries installed in their customers' homes to offer balancing services to stabilize the electrical grid. Tesla's VPP was able to aggregate more than 4,500 customers with photovoltaic systems and Powerwall batteries, offering flexibility services to the Californian system operator for powers of more than 30 MW, (Tes). Following the California experience, the "Energy Efficiency Summer Reliability Program" was launched in the United States, allowing DERs to offer balancing services to the grid. Following the launch of the SUNRUN program, a company that sells and installs domestic photovoltaic and battery solutions started a VPP in the portion of the electrical grid operated by Independent System Operator - New England (ISO-NE), which promises to aggregate and coordinate more than 7,500 residential home solar and battery systems, creating a virtual power plant capable of discharging 30 megawatts of clean energy back to the grid, (Sun).

In Europe, most states allow aggregated DERs to participate in the balancing services market. A complete map of the states that allow participation and the constraints they impose is provided by Smart Energy Europe (SMARTEU). Companies that sell and operate VPPs include NextKraftwerke that operates in seven european country and manage portfolios with more of 2500 MW of DERs, (Kra).

4. The fading of network perimeter and novel cybersecurity issues

4.1. The change in network architecture

Due to this change of paradigm, the methods and technologies that are necessary to protect these infrastructures from cyberattack must completely evolve. Microgrids are usually based on traditional SCADA systems. Cybersecurity of SCADA systems has been broadly investigated in the last past years, and has strong foundation (Colbert and Kott (2016)). With the advances in the sophistication of attacks, research is still ongoing. Yadav and Paul (2021) present a review of architecture and security of SCADA systems, discussing the evolution from monolithic systems to IoT-based systems, a taxonomy of the attacks, and comparative analysis of state-of-the-art IDSs and SCADA testbeds. Also, an important issue is how to decline cybersecurity countermeasures to specific domains of applications, respecting its peculiar operational requirements. Gaggero et al. (2021) discuss the latest advancement in microgrid cybersecurity research, identifying 5 main fields: the application of encryption and authentication techniques in common industrial protocol, that has been standardized in the IEC 62351; the use of Software Defined Networking in order to improve the resiliency of the control system by dynamically re-configuring the network after the detection of faults and/or cyberattacks; the advancement in Host-Intrusion Detection Systems for real time operating systems and time-critical devices in the industrial environment; the physics-based anomaly detection approach for cybersecurity monitoring in the industrial field; the resilient control strategies that can be implemented in DERs, especially in the case the microgrid has to operate in islanded mode.

The core concept for designing SCADA systems is segmentation (Stouffer et al. (2011)). The aim of network segmentation is to minimize the possibility to

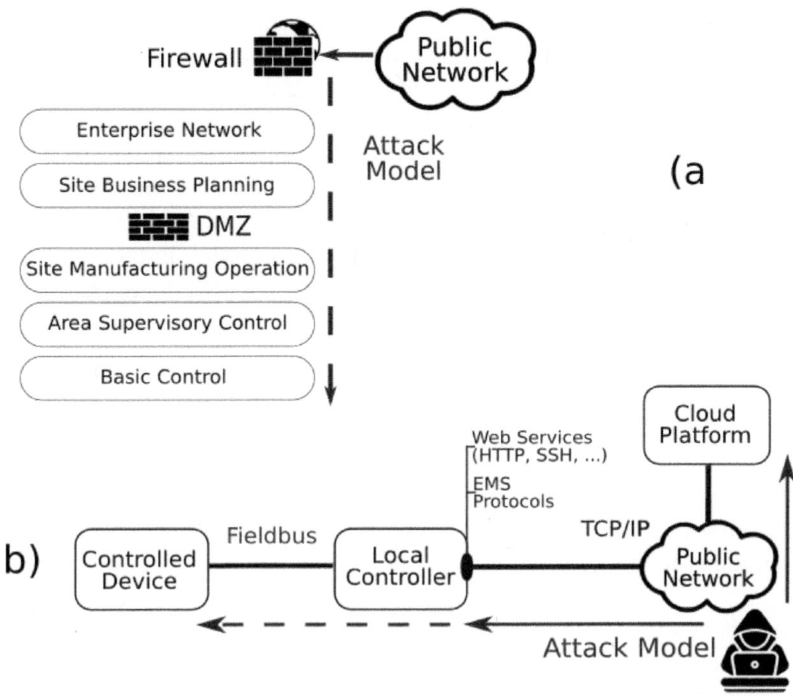

Figure 2. The evolution of control networks in smart distribution systems.

access to sensitive information for those systems and people who don't need it by using different techniques and technologies depending on the network's architecture and configuration. In this sense, a SCADA system must be designed with clear boundaries and points of access to/from external networks, and always protected through firewalls and other technologies for perimeter defense. A classical design reference has been the so called "Purdue model" (see Figure 2a). In this model, the network is divided in multiple zones, basing on the logical proximity to physical sensors and actuators. The threat model is represented by an attacker that, coming from the public internet, has the aim to modify the behavior of the control system acting on the lower layers of the architecture, and in order to do that, it has to violate multiple lines of defense. The defense strategies are designed accordingly: the defense-in-depth paradigm prescribe to design the defense in multiple lines (Fabro et al. (2016)), including both organizational and technical aspect, starting from the defense of the perimeter of the newtork by using technologies such as firewalls and de-militarized zones, to security monitoring.

Nevertheless, as discussed in the previous section, the power systems, and in particular the distribution systems, are increasingly using the public communication network for monitoring and control purposes. In the previously discussed technologies, EMS are located in a cloud platform, that communicates with the field devices through different technologies, but without relying on dedicated infrastructures. We therefore must face the issue to protect systems whose architecture is similar to the one shown in Figure 2b. In this scenario, part of the communication is done by using a public network. The platform is a web server, and also the local controller is a web server, that can be hidden within a private network or directly have a public IP address. In this context, the Purdue model is not applicable anymore. If the control system rely on the public network, the concept of segmentation completely falls. The attack model, in this case, is represented by an attacker that can communicated directly with the platform and/or with the local controller.

4.2. Attack surface

As discussed in the previous section, smart distribution systems act as a SCADA: in facts, similarly to a SCADA system for a microgrid, Smart Distribution Systems make use of a telecommunication network to send commands such as the power setpoints for DERS; the main difference is that Smart Distribution Systems rely on the public network, so that the concepts of segmentation and segregation completely fall. In this scenario, it is necessary to protect these systems taking for granted that an attacker is able to reach the machines by a remote attack. We can identify three main attack models:

- Attacks toward the platform.
- Attacks toward the communication stack.
- Attacks toward the local controller.

The platform is typically a web server, that can expose different types of services, including specific services for the communication with the local controller, but also common protocols such as HTTP or SSH for remote control and maintenance or for allowing users to access specific services. The platform, therefore, can suffer from common vulnerabilities of we services, such as SQL injection, cross-site scripting, broken authentication and session management and Denial of Services. The risk associated to this vulnerabilities is very high since, as we will discuss in the next section, the control of the platform could lead also to the failure of the grid.

The same vulnerabilities can be found on the local controller, but the scenario is different. In this case, while the associated risk is significantly lower, since the attack could lead to the control of a single user per time, it is worth noticing that the local controller may consist of cheap hardware with limited computational capabilities. In this case, it is more likely that the devices contains vulnerabilities, and moreover it is difficult to implement common cybersecurity countermeasures: host-intrusion detection systems are usually heavy software, that could be hardly be implemented in these devices.

Also, the communication stack could be prone to cyber attacks. In particular, there is not yet a common standard for the application-layer protocol between the local controller and the platform. Taking, for example, the case of electric vehicle charging, the solution currently employed include OCPP (Open Charge Point Protocol), OSCP (Open Smart Charging Protocol), ISO 15118, OpenADR, but also general-purpose protocols such as MQTT, and many more, while each one of them has different versions. In this scenario, different vulnerabilities in the communication stack could be present and should be further investigated.

We can find the first evidences of this change of paradigm in the literature. Skarga-Bandurova et al. (2022) discuss innovative elements in the smart city from a cybersecurity perspective, with a particular focus on electric vehicle charging, pointing out that not all solutions have adequate cybersecurity protection. Nasr et al. (2022) provide a significant example of the evaluation of the cyber threats in the context of Electric Vehicle Charging Systems; first, authors demonstrate the feasibility of cyber attacks against the deployed EVCS, such as such as SQL Injection (SQLi), Cross-Site scripting (XSS), Server-Side Request Forgery (SSRF), and Cross-Site Request Forgery (CSRF) showing how this can lead to remote EVCS exploitation and manipulation; then, they simulate the attack on the power system, showing hot these may cause frequency instability, which results in possible power outages and/or denial of service. Gaggero et al. (2023) specifically take into account the evolution of energy communities, evaluating the technologies that are employed, the attack surface, and also discusses some possible research directions to address these issues. Some works analyze risk associated to Virtual Power Plants. Venkatachary et al. (2021) present a Edge-based security architecture to secure VPPs and ensure privacy and data protection so to reduce the risks. Khan et al. (2021) presents a cybersecurity analytics system for the detection of a cyber-attacker manipulating the VPP cyber layer operation set- points gradually to violate network stability bound. Li et al. (2018) proposes a distributed economic dispatch strategy of the VPP that is robust against cyber-attacks (i.e., noncolluding and colluding attacks) and resilient to communication failures. Despite evident similarities, research is at an early stage , and there is not yet evidence in the literature regarding attacks toward all the technologies that we discussed in Section 3, and further work has to be done in this field.

4.3. Potential impact of the attacks

Depending on the technology, a single EMS platform may control a huge number of generators and/or controllable loads in the distribution grid. Moreover, these devices may have some peculiar features in common, such as in the case of energy communities, in which the devices are placed under a single substation. In case of a successful cyber attack, the attacker may be able to control a large number of devices in a malicious manner. While, from one side, this may cause economic damages, more relevantly it is possible that such an attack may cause malfunctioning of the power grid. Bhattarai et al. (2018) present a cyber attack scenario targeting DERS, in which the power output of the DER is manipulated to cause sustained oscillations or even system instability. Authors still do not provide a deep evaluation of the locations where DER can be manipulated to produce a higher impact. Tuttle et al. (2019) also provide an evaluation of the impact of the control of a large number of DERs, but focusing on storage systems. Linnartz et al. (2021) show how it is possibile to violate voltage boundaries through a cyberattacks on DERs on the CIGRE medium voltage benchmark grid. Nevertheless, few works perform simulations by considering the features and the constraint of a single technology for the control of a large number of DERs. Nasr et al. (2022) take into account the case of electric vehicle charging for assessing the impact of the cyber attack on that infrastructure. Still, the evaluation of the impact of cyber attacks on the distribution grid has to be further investigated. In particular, researchers should take into account specific use cases of the technologies discusses in the previous section. For example, in the case of energy communities, a single platform may control at maximum 1 MW of generation under a single substation, but also a certain amount of controllable loads, depending on the national laws. This evaluation is extremely important to assess the risks of such technologies.

5. A framework for cybersecurity monitoring

The traditional approach for cybersecurity monitoring in SCADA systems can not be applied anymore in the smart distribution grid. It is therefore necessary to develop new procedures and guidelines. We propose a framework that can represent a first step for developing a proper cybersecurity monitoring scheme for companies that owns and manages such infrastructures. An overall scheme is shown in Figure 3.

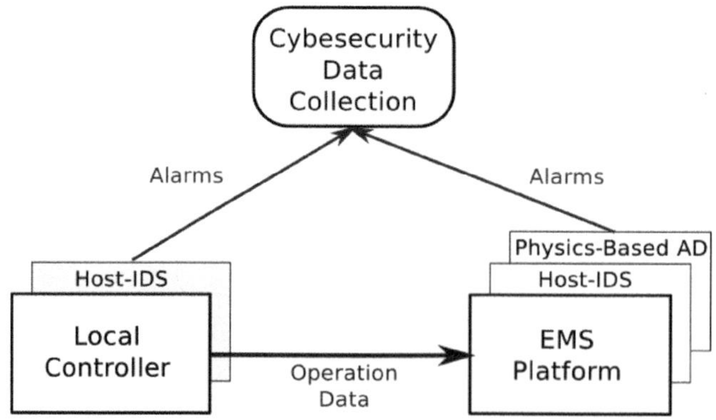

Figure 3. Framework for cybersecurity monitoring in smart distribution systems.

The first step is to define how to collect and analyze cybersecurity data. Energy utilities must develop organizational procedures to properly monitor these technologies. In this case, the major issue could be represented by the huge number of systems managed by a single company. A single energy utility may easily control tens or hundreds of platforms for smart distribution systems, each one of these are composed of hundreds of local controller distributed over a wide area. It is evident that the process of cybersecurity monitoring must be deeply automatized, and rely on human supervision only after a strong pre-processing of data. Usually energy utilities make use of Security Information and Event Management Systems (SIEMs) to collect data from different devices, including operational technology elements, such as firewalls, network intrusion detection systems etc, to store and analyze that data, correlating data sources. A usual choice is to centralize the collection of logs which come from different OT sources in the same SIEM platform. Usual OT data sources for energy utilities include firewalls, probes and intrusion detection systems dedicated to OT traffic and control servers, as well as cloud services. Nevertheless, developing correlation rules for the integration of IT and OT data sources is not a trivial task. At the current stage, the integration of data coming from the technologies discussed above into the company SIEM is an open issue.

After the definition of a platform for collecting cybersecurity data, allowing proper visibility of the plants, it is necessary to deploy systems able to detect incoming attacks from the field. The two devices mainly exposed to cyberattacks are the EMS platform, and the local controller. The scenario for these two devices is totally different.

EMS platforms inherit the well-known cybersecurity issues of web servers and cloud computing; at the same time, the technology for the monitoring of web servers is mature. There are different commercial solutions of Intrusion Detection Systems for these kinds of systems. In particular, technologies include Endpoint Detection and Response (EDR) and Extended Detection and Response (XDR) systems. Endpoint protection systems can collect security telemetry from endpoints, cloud workloads, network email, and more. Endpoint protection can be used as a log source by SIEMs.

Local controllers, on the contrary, are usually devices with limited computational resources. In this case, the challenge is to develop IDS specifically developed for the devices. We consider two main examples: an electric vehicle charging station (EVCS), and a smart gateway for an user of an energy community (Gaggero et al. (2023)). An EVCS may expose different web services, such as HTTP, SSH and so on, for remote monitoring and maintenance purposes, in addition to the communication with the EMS. The challenge here is twofold: an IDS must take into account the specific features of the communication of the EVCS, and its constraints. Some papers in the literature are starting to take into account this issue. For example, ElKashlan et al. (2023) propose a machine learning-based intrusion detection system for IoT electric vehicle charging stations; while the algorithm shows good performances, the work still lacks of the implementation details in the network architecture. Further work has to be done in this field. For smart gateways in energy communities, the problem is even more serious, since the computational constraints of these devices are severe. Much effort has to be done for the development of proper IDS solutions in this field.

The only monitoring of network parameters could not be enough to monitor these systems properly. Another fundamental difference between traditional SCADA and smart distribution systems is the number of deployed systems. An energy utility, even a bigger one, could manage tens, or at most thousands of generation plants of big dimensions. SCADA systems are usually monitored most of the time by human operators, that have the role to monitor the physical parameters of the process, ensuring that the process is working in a safe operational mode. For smart distribution systems, the number of generators increases exponentially. The number of photovoltaic systems, storage systems EVCS, and controllable loads that refers to one of the platforms managed by a single utility may easily reach the order of magnitude of thousands. In this sensor, the monitoring must be automatized. An innovative field of research is represented by physics-based anomaly detection (Giraldo et al. (2018)); in this case, the objective is to notice anomalies in the process directly taking into account the physical measures extracted from the process itself. Usually based on machine learning and deep learning technologies, physics-based anomaly detection algorithms utilize features such as, in the case of the electrical field, voltages, currents, frequencies, and powers, instead of network parameters. For example, Gaggero et al. (2020) and Gaggero et al. (2022) propose an anomaly detection algorithm for a photovoltaic system and a battery storage system, that makes use of a neural network architecture called autoencoder to directly analyze the parameters measured and sent to a higher level controller from the inverter. These kinds of algorithms could be implemented within the EMS platforms, that already receive the physical measures for general-purpose monitoring and control functions. The idea is to exploit the measures that the EMS already manages to also detect potential anomalies related to cyberattacks. Then the alarms can be sent to the platform that collects cybersecurity data, which can be further elaborates the alarm with specifically developed correlation rules. An overall summary of log sources that can be extracted from control devices and sent to a centralized SIEM platform can be found in Table 1.

The proposed approach mimic the monitoring functions that are present in a SCADA system: the Intrusion Detection Systems have to be adapted to respect the network architecture and the computational constraints, while the physical monitoring is shifted from human supervision to automatic approaches.

Table 1.   Summary of SIEM Log sources.

SIEM Log Sources	
Cloud Platforms	• EDR • XDR • Raw Logs (Authentication, ...) • Physics-Based Intrusion Detection
Local Controllers	• Lightweigh Host-IDS • Raw Logs (Authentication, ...)

The effectiveness of the proposed approach can hardly be demonstrated by using simulation environments: in fact, very few data related to attacks in the real world are available, since these infrastructures are in an early stage. Still, some considerations in respect to the existing standards for cybersecurity of industrial control systems can be done. The main standards developed in the sector regarding cybersecurity are the ISA/IEC-62443 and the NIST SP 800-82. These standards recommend the use of cybersecurity monitoring tools to properly protect these systems. The deployment of security monitoring tools usually follows three main phases:

(1) Definition of security requirement: in this phase, the company logically separates all the control systems into different zones, and assigns a Security Level to each of them that is a number correlated to the skills level of the malicious actors' and the consequent complexity of the cyber-attack.

(2) Definition of requirements satisfied by tools: in the phase of assessing cybersecurity monitoring tools and their adherence to standards, various approaches can be explored to meet the requirements and achieve a specific Security Level.

(3) Deployment of Security Monitoring Tools: finally, the tools that satisfy the requirements of a specific Security Level and at the same time respect the operational constraints can be implemented.

Depending on the specific network architecture, the proposed framework plays a crucial role to reach all the security requirements for a Security Level 2 or 3. Examples of requirements to reach a Security Level 2 are *SR 2.8 – Auditable Event, SR 3.3 – Security functionality verification, SR 6.1 – Audit log accessibility*, that can be reached by obtaining a proper visibility of the smart distribution system on the SIEM, reducing the complexity of the analysis by using appropriate tools such as EDR on the cloud platform and Lightweigh Host-IDSs on the local controllers. This is coherent with the criticality that an attack towards these systems might represent for the whole power systems. Nevertheless, further work has to be done to properly adapt these considerations to the variety of network architectures that are arising with the evolution of smart distribution systems.

6. Conclusions

This paper analyzed the evolution of power distribution systems, discussing novel cybersecurity issues that are arising due to the employment of innovative technologies for monitoring and control purposes. The paper highlights some open issues in the state of the art: in particular, further work has to be done to simulate the effects of cyberattacks in emerging technologies such as energy communities and virtual power plants, and how to deal with the risk in a scenario in which the distribution system operator may not have the control of a large number of devices, managed instead by third-party entities. Also, this paper proposed a preliminary framework to design a proper cybersecurity monitoring strategy for energy utilities. This framework involves the deployment of monitoring tools such as Endpoint Protection and/or Physics-based Intrusion Detection, and the integration with event management platforms such as SIEMs. Still, further work has to be done to clearly define the data needed by the SIEM platform to have sufficient visibility of a wide range of attacks towards these systems.

Funding

This work was partially supported by project RAISE under the MUR National Recovery and Resilience Plan funded by the European Union - NextGenerationEU.

References

Bhattarai, R., Hossain, S. J., Qi, J., Wang, J. and Kamalasadan, S. 2018. Sustained system oscillation by malicious cyber attacks on distributed energy resources. In 2018 IEEE Power & Energy Society General Meeting (PESGM), pp. 1–5.

Chudoba, C. and Borges, T. 2022. Platform-based energy communities in Germany and their benefits and challenges. In Energy Communities, pp. 385–397. Elsevier.

Colbert, E. J. and Kott, A. 2016. Cyber-security of scada and other industrial control systems (Vol. 66). Springer.

Cybersecurity for distributed energy resources and smart inverters. n.d. https://www.osti.gov/servlets/purl/1374586 (Accessed: 2022-05-01).

ElKashlan, M., Elsayed, M. S., Jurcut, A. D. and Azer, M. 2023. A machine learning-based intrusion detection system for iot electric vehicle charging stations (EVCSS). Electronics, 12(4): 1044.

El Mrabet, Z., Kaabouch, N., El Ghazi, H. and El Ghazi, H. 2018. Cyber-security in smart grid: Survey and challenges. Computers & Electrical Engineering, 67: 469–482.

Fabro, M., Gorski, E., Spiers, N., Diedrich, J. and Kuipers, D. 2016. Recommended practice: Improving industrial control system cybersecurity with defense-in-depth strategies. DHS Industrial Control Systems Cyber Emergency Response Team.

Flexo - community manager - our platform for managing and optimising energy communities. n.d. https://www.hivepower.tech/flexo/flexo-community-manager (Accessed: 2022-10-30).

Gaggero, G. B., Caviglia, R., Armellin, A., Rossi, M., Girdinio, P. and Marchese, M. 2022. Detecting cyberattacks on electrical storage systems through neural network based anomaly detection algorithm. Sensors, 22(10): 3933.

Gaggero, G. B., Girdinio, P. and Marchese, M. 2021. Advancements and research trends in microgrids cybersecurity. Applied Sciences, 11(16): 7363.

Gaggero, G. B., Piserà, D., Girdinio, P., Silvestro, F. and Marchese, M. 2023. Novel cyber-security issues in smart energy communities. In 2023 1st International Conference on Advanced Innovations in smart cities (ICAISC), pp. 1–6.

Gaggero, G. B., Rossi, M., Girdinio, P. and Marchese, M. 2020. Detecting system fault/cyberattack within a photovoltaic system connected to the grid: A neural network-based solution. Journal of Sensor and Actuator Networks, 9(2): 20.

Giraldo, J., Urbina, D., Cardenas, A., Valente, J., Faisal, M., Ruths, J., . . . Candell, R. 2018. A survey of physics-based attack detection in cyber-physical systems. ACM Computing Surveys (CSUR), 51(4): 1–36.

González-Granadillo, G., González-Zarzosa, S. and Diaz, R. 2021. Security information and event management (siem): Analysis, trends, and usage in critical infrastructures. Sensors, 21(14): 4759.

Justo, J. J., Mwasilu, F., Lee, J. and Jung, J. -W. 2013. Ac-microgrids versus dc-microgrids with distributed energy resources: A review. Renewable and Sustainable Energy Reviews, 24: 387–405.

Khan, A., Hosseinzadehtaher, M., Shadmand, M. B. and Mazumder, S. K. 2021. Cyber-security analytics for virtual power plants. In 2021 IEEE 12th International Symposium on Power Electronics for Distributed Generation Systems (PEDG), pp. 1–5.

Li, P., Liu, Y., Xin, H. and Jiang, X. 2018. A robust distributed economic dispatch strategy of virtual power plant under cyber-attacks. IEEE Transactions on Industrial Informatics, 14(10): 4343–4352.

Linnartz, P., Winkens, A. and Simon, S. 2021. A method for assessing the impact of cyber attacks manipulating distributed energy resources on stable power system operation. In 2021 IEEE PES Innovative Smart Grid Technologies Europe (ISGT Europe), pp. 01–05.

Nasr, T., Torabi, S., Bou-Harb, E., Fachkha, C. and Assi, C. 2022. Power jacking your station: In-depth security analysis of electric vehicle charging station management systems. Computers & Security, 112: 102511.

(NESCOR), N. E. S. C. O. R. 2013. Cyber security for der systems.

Pereira, H., Gomes, L., Faria, P., Vale, Z. and Coelho, C. 2021. Web-based platform for the management of citizen energy communities and their members. Energy Informatics, 4(2): 1–17.

Pourghaderi, N., Fotuhi-Firuzabad, M., Moeini-Aghtaie, M. and Kabirifar, M. 2018. Commercial demand response programs in bidding of a technical virtual power plant. IEEE Transactions on Industrial Informatics, 14(11): 5100–5111.

Pudjianto, D., Ramsay, C. and Strbac, G. 2007. Virtual power plant and system integration of distributed energy resources. IET Renewable Power Generation, 1(1): 10–16.

Radoglou-Grammatikis, P., Sarigiannidis, P., Iturbe, E., Rios, E., Martinez, S., Sarigiannidis, A., . . . others. 2021. Spear SIEM: A security information and event management system for the smart grid. Computer Networks, 193: 108008.

Radoglou-Grammatikis, P. I. and Sarigiannidis, P. G. 2019. Securing the smart grid: A comprehensive compilation of intrusion detection and prevention systems. IEEE Access, 7: 46595–46620.

Rose - un software per la creazione e la gestione completa delle comunità energetiche. n.d. https://solutions.mapsgroup.it/comunita-energetiche (Accessed: 2022-10-30).

Rouzbahani, H. M., Karimipour, H. and Lei, L. 2021. A review on virtual power plant for energy management. Sustainable Energy Technologies and Assessments, 47: 101370.

Skarga-Bandurova, I., Kotsiuba, I. and Biloborodova, T. 2022. Cyber security of electric vehicle charging infrastructure: Open issues and recommendations. In 2022 IEEE International Conference on Big Data (Big Data), pp. 3099–3106.

Stouffer, K., Falco, J., Scarfone, K. et al. 2011. Guide to industrial control systems (ICS) security. NIST Special Publication, 800(82): 16–16.

Sunrun and PGE collaborate on residential battery-powered virtual power plant to support grid reliability for electric customers. n.d. https://investors.sunrun.com/news-events/press-releases/detail/279/sunrun-and-pge-collaborate-(Accessed: 2023-04-30).

Tesla virtual power plant (beta). n.d.. https://www.tesla.com/support/energy/powerwall/own/california-virtual-(Accessed: 2023-04-30).

The power of many. (n.d.). https://www.next-kraftwerke.com/ (Accessed: 2023-04-30).

Tuttle, M., Poshtan, M., Taufik, T. and Callenes, J. 2019. Impact of cyber-attacks on power grids with distributed energy storage systems. In 2019 IEEE International Conference on Communications, Control, and Computing Technologies for Smart Grids (smartgridcomm), pp. 1–6.

Tuyen, N. D., Quan, N. S., Linh, V. B., Vu, T. V. and Fujita, G. 2022. A comprehensive review of cybersecurity in inverter-based smart power system amid the boom of renewable energy. IEEE Access.

Venkatachary, S. K., Alagappan, A. and Andrews, L. J. B. 2021. Cybersecurity challenges in energy sector (virtual power plants)-can edge computing principles be applied to enhance security? Energy Informatics, 4(1): 5.

Wang, W. and Lu, Z. 2013. Cyber security in the smart grid: Survey and challenges. Computer Networks, 57(5): 1344–1371.

Yadav, G. and Paul, K. 2021. Architecture and security of scada systems: A review. International Journal of Critical Infrastructure Protection, 34: 100433.

CHAPTER 6

An Overview of Artificial Intelligence Driven Li-Ion Battery State Estimation

Mohammed Hindawi,[a] *Yasir Basheer*,[b] *Saeed Mian Qaisar* [a,c,]*
and *Asad Waqar*[d]

1. Introduction

1.1. Context of the study

Battery state estimation is a critical process that involves assessing the current and future state of a battery, typically in terms of its capacity, voltage, and overall health. It plays a crucial role in various domains, such as electric vehicles, portable electronics, and renewable energy systems (Yang et al., 2023). Strictly speaking, the problem of Li-ion battery state estimation involves accurately estimating the state of charge (SoC), state of health (SoH), Capacity, Remaining Useful Life (RUL), and other critical parameters of the battery based on avail-

[a] CESI LINEACT, Lyon, 69100, France.

[b] Department of Electrical Engineering, Riphah International University, Islamabad 46000, Pakistan.

[c] Electrical and Computer Engineering Department, Effat University, Jeddah, 22332, Saudi Arabia.

[d] Department of Electrical Engineering, Bahria School of Engineering and Applied Sciences (BSEAS), Bahria University, Islamabad 44000, Pakistan.

* Corresponding author: sqaisar@effatuniversity.edu.sa

able measurements (Hannan et al., 2017; Qaisar, 2020; Qaisar and AbdelGawad, 2021). The problem of Artificial Intelligence (AI)-driven battery state estimation involves developing an AI model that can accurately estimate the state of a battery based on input data (Krichen et al., 2023; Ren and Du, 2023). The objective is to capture the complex relationships between the battery's behavior and the collected data, while accounting for various operating conditions and factors such as temperature and aging effects. Artificial intelligence (AI) techniques have emerged as powerful tools for addressing this problem and improving the accuracy and reliability of Li-ion battery state estimation (Galiounas et al., 2022; Mian Qaisar et al., 2021). The application of AI techniques, including deep learning, support vector machines, and incremental capacity analysis demonstrates the potential of AI-driven methods to accurately estimate the state of Li-ion batteries, improving their performance, management, and overall utilization.

Smart grids and batteries are integral components of modern energy systems, offering numerous advantages for electricity generation, distribution, and consumption. Smart grids employ advanced technologies to enhance efficiency, reliability, and sustainability (Gungor et al., 2011; Qaisar and AlQathami, 2021). They facilitate real-time monitoring, control, and optimization of energy flow, enabling effective integration of renewable energy sources and demand response programs. Batteries, in turn, play a crucial role in the development and operation of smart grids, offering significant benefits and addressing key challenges in the modern electricity landscape (Roberts and Sandberg, 2011). One of the primary advantages of batteries in smart grids is their ability to store and release electricity on demand. As intermittent renewable energy sources like solar and wind become increasingly prevalent, batteries provide a means to capture excess energy generated during peak production and discharge it when demand is high or renewable generation is low. Additionally, batteries enable the integration of distributed energy resources, such as rooftop solar panels and small-scale wind turbines, by absorbing and dispatching their output as needed. By facilitating a more balanced and reliable energy supply, batteries contribute to the overall resilience and flexibility of the smart grid (Alirezazadeh et al., 2021). Furthermore, batteries offer ancillary services, including frequency regulation and voltage support, which enhance grid stability and improve power quality (Kundur and Malik, 2022).

Battery technology has undergone significant advancements in recent years, revolutionizing the way user's power their devices and vehicles. Traditional battery technologies, such as lead-acid and nickel-cadmium, have been largely replaced by more efficient and environmentally friendly alternatives. One notable breakthrough is the development of lithium-ion batteries. These compact and lightweight batteries offer high energy density, allowing them to store and deliver more power in a smaller package (Hassan, 2023). Lithium-ion batteries have become the preferred choice for portable electronics, electric vehicles, and renewable energy systems. Researchers are continuously working on improving lithium-ion batteries through various approaches. For example, the incorporation of silicon-based anodes has shown promise in enhancing the energy storage capacity of lithium-ion batteries (Ali et al., 2023b). Another promising area of battery technology is the development of solid-state batteries. These batteries utilize solid electrolytes, eliminating the need for flammable liquid electrolytes and improving safety (Bachman et al., 2016). Solid-state batteries also offer the potential for higher energy density and wider temperature ranges. Additionally, the exploration of alternative battery chemistries, such as lithium-sulfur batteries and lithium-air batteries, holds promise for even higher energy densities and lower (Bruce et al., 2011). Battery technology also involves optimizing charging infrastructure and exploring innovative manufacturing techniques, such as 3D printing, to enhance efficiency and reduce costs (Lyu et al., 2021). Moreover, recycling and second-life applications aim to mitigate environmental concerns and extend battery usefulness (Iqbal et al., 2023).

Batteries have become essential technology with a wide range of applications in various industries. In the field of portable electronics, lithium-ion batteries have revolutionized the market by providing high energy density and long-lasting power (Gandoman et al., 2021). Battery energy storage systems (BESS) (Ali et al., 2023a) find applications in renewable energy systems, enabling the storage of excess energy during periods of low demand and its release during high demand, thus enhancing grid stability and facilitating the integration of intermittent renewable sources. Electric vehicles (EVs) heavily rely on batteries for their propulsion (Gandoman et al., 2022), with lithium-ion batteries being the dominant choice due to their high energy density, improved range, and decreasing costs. Batteries also find applications in off-grid and remote areas, providing reliable power where traditional grid infrastructure is not available (Aldosary et al., 2021). Furthermore, batteries serve as backup power systems in critical infrastructure, ensuring uninterrupted operation during power outages (Quynh et al., 2021). Grid-scale energy storage systems based on batteries are being deployed to manage peak demand, optimize load balancing, and support grid stability (Chen et al., 2020). Moreover, batteries are used in aerospace applications, including satellites, space probes, and aircraft, where lightweight, high-performance batteries are essential for power storage (Kühnelt et al., 2022). The versatile applications of batteries continue to expand as technology advances, driving innovation and sustainable energy solutions across industries, while providing reliable and portable power sources for a variety of purposes (Qaisar and Alyamani, 2022).

1.2. Advances in battery state estimation due to AI

Advances in battery state estimation have been significantly enhanced by the application of Artificial Intelligence (AI) techniques (Zhang et al., 2022). Here, the authors delve into the key advances and benefits brought about by the integration of AI in battery state estimation.

Enhanced Accuracy: AI algorithms, such as artificial neural networks (ANNs) and support vector machines (SVMs), can capture complex non-linear relationships within battery systems. These algorithms learn from vast amounts of training data and can model the intricate electrochemical processes occurring inside the battery. As a result, AI-driven battery state estimation can provide more accurate predictions of SoC and SoH compared to traditional methods (Ren and Du, 2023).

Adaptability to Varying Conditions: Batteries experience dynamic operating conditions, such as temperature variations, load changes, and aging effects. AI algorithms can adapt to these varying conditions and provide accurate state estimation in real-time. The ability of AI models to continuously learn and update themselves using new data enables adaptive state estimation that remains effective over the battery's lifetime (Zhang et al., 2019).

Improved Robustness: Traditional battery state estimation methods often struggle to handle extreme conditions and complex behaviors. AI-driven approaches can handle these challenges more effectively. For example, AI models can account for non-linear relationships between battery voltage, current, and SoC/SoH, as well as handle scenarios with partial observability or missing data (Shu et al., 2021).

Fast and Real-Time Estimation: AI algorithms can process large amounts of data and make predictions quickly, enabling real-time battery state estimation. This capability is crucial in applications that require immediate and accurate information about the battery's state, such as electric vehicles or critical power systems. Real-time estimation allows for better control and optimization of battery operation, leading to improved performance and extended battery life (Patil, 2023).

Data-Driven Insights: AI-driven battery state estimation generates insights from vast amounts of data collected during operation. These insights can be used to gain a deeper understanding of battery behavior, identify degradation patterns, optimize battery usage, and develop advanced control strategies (Ling, 2022). The data-driven approach enables predictive maintenance, where early signs of battery degradation can be detected and addressed proactively.

Reduced Dependence on Battery Models: Traditional battery state estimation methods often rely on accurate battery models, which can be challenging to develop due to the complexity of battery behavior. AI-driven approaches are model-free or model-light, as they learn directly from data. This reduces the dependency on precise battery models and makes the estimation process more versatile and adaptable to different battery chemistries and configurations (Singh et al., 2023).

Integration with Sensor Fusion: AI-driven battery state estimation can integrate data from multiple sensors, such as voltage, current, temperature, and impedance, to achieve more comprehensive and accurate state estimation. Sensor fusion techniques, combined with AI algorithms, enable a holistic understanding of battery behavior by incorporating different sources of information (Schneider and Endisch, 2020).

The advances in AI-driven battery state estimation have significant implications for various industries, including electric vehicles, renewable energy systems, and portable electronics. Accurate and real-time estimation of battery state enables optimal utilization, prolongs battery life, enhances safety, and contributes to the overall efficiency and sustainability of energy storage systems. Continued advancements in AI algorithms, coupled with the availability of large-scale battery datasets, will further enhance the capabilities and performance of AI-driven battery state estimation techniques.

1.3. Use of AI assistive battery state estimation in Battery Management System

The integration of AI-assisted battery state estimation in the Battery Management System (BMS) has significant implications for various applications. In electric vehicles (EVs), accurate estimation of battery states enhances range prediction and enables efficient power management, optimizing vehicle performance and user experience (Liu et al., 2022). In renewable energy systems, AI-assisted BMS ensures effective energy utilization, improves grid stability, and extends battery lifespan (Hu et al., 2019). In portable electronics, intelligent BMS enables precise SoC estimation, enhancing battery runtime and user satisfaction (Buchmann, 2001). In details, AI-driven battery state estimation has the potential to improve the accuracy, robustness, and real-time capabilities of BMSs (Raoofi and Yildiz, 2023). These techniques utilize data-driven models and patterns to estimate critical battery states, including SoC, SoH, and state of power (SoP) (Hossain Lipu et al., 2022). AI-based state estimation methods have shown great potential in overcoming challenges associated with nonlinearities, and environmental variations in battery systems.

Machine learning algorithms, such as support vector machines (SVM), random forests, and neural networks, have been widely adopted for battery state estimation (Manoharan et al., 2022). SVM algorithms have been utilized to estimate battery SoC based on measured voltage and current data (Álvarez Antón et al., 2013). Random forests have demonstrated accuracy in estimating SoH using battery impedance measurements (Li et al., 2018b). Furthermore, neural networks, particularly long short-term memory (LSTM) networks, have been successful in predicting battery performance and degradation (Li et al., 2019b). Furthermore, Deep learning techniques have shown remarkable progress in battery state estimation due to their ability to extract complex patterns from large datasets (Zhang et al., 2022). Convolutional neural networks (CNNs) have been applied to estimate SoC based on battery voltage profiles (Fan et al., 2022). Recurrent neural networks (RNNs) and their variants, such as gated recurrent units (GRUs) and LSTMs, have been employed for accurate SoH prediction and remaining useful life (RUL) estimation (Song et al., 2018). Generative adversarial networks (GANs) have been utilized to model battery degradation processes and predict SoH degradation trajectories (Yang et al., 2022).

The integration of AI techniques with other advanced tools, such as system identification, Bayesian inference, and particle filters, has further improved battery state estimation (Saha et al., 2007). System identification methods combined with AI algorithms enable the identification of battery models and parameter estimation for accurate state estimation. Bayesian inference approaches provide probabilistic estimation of battery states, offering uncertainty quantification in the estimation process (Saha et al., 2007). Particle filters, such as the unscented Kalman filter (UKF) and the extended Kalman filter (EKF), have been employed for real-time battery state estimation by incorporating AI-based models (Konatowski et al., 2016). These AI-assisted battery state estimation techniques in BMS provide numerous benefits, including increased safety, enhanced battery performance, and optimized energy management.

1.4. Contributions

This study focuses on the critical process of battery state estimation, specifically in terms of capacity, voltage, and overall health, and highlights how the artificial intelligence (AI) techniques enhance the battery state estimation mechanism. The uniqueness of this work lies in the integration of AI-assisted battery state estimation in the Battery Management System (BMS). It offers benefits such as increased safety, enhanced battery performance, and optimized energy management. The primary objective of this study is to provide a comprehensive overview of AI-driven battery state estimation workflows, including data collection and preprocessing, feature engineering, battery model construction, and application. By incorporating AI techniques, the study aims to improve the accuracy, adaptability, and robustness of battery state estimation, leading to optimal utilization and extended battery life.

2. Literature review

2.1. Battery technology

There are different types of batteries with various applications. Batteries can be either primary or secondary batteries. Primary batteries are disposable, while secondaries are rechargeable. Rechargeable batteries, also known as storage batteries, accumulators, or Secondary batteries are batteries that have the capability to undergo multiple cycles of charging, discharging into a load, and subsequent recharging. This is opposed to one-time use batteries also known as the primary batteries that are used and then disposed. Secondary batteries come in different types. The subsequent are frequent examples of secondary or rechargeable batteries (Sarmah et al., 2023).

Lead acid batteries, which use acids and lead as electrodes to store energy. One notable benefit is their capacity to deliver substantial surge currents, coupled with their affordability. They are mostly used as sources for power for starter motors. However, they have a low energy to weight ratio making them unsuitable for many tasks (Mansuroglu et al., 2023).

Nickel cadmium batteries use Nickel Oxide hydroxide and metallic cadmium. These batteries are not environmentally friendly since Cadmium is a toxic element, and they have been banned in most countries (Rana et al., 2023).

Nickel-Metal Hydride Batteries use hydrogen-absorbing alloy instead of the toxic Cadmium electrode. These are now common in most industrial usage and in consumer types (Krishnamoorthy et al., 2023). However, their energy to volume ratio is also high making them not applicable in many areas.

Lithium-Ion Batteries Although there are different types, lithium cells are one of the most used cell types today. Some of the reasons why lithium systems are preferred are, they have a higher energy density, they have a superior cold temperature performance. They possess an extended period of effectiveness and offer cost efficiency, as demonstrated in Figure 1.

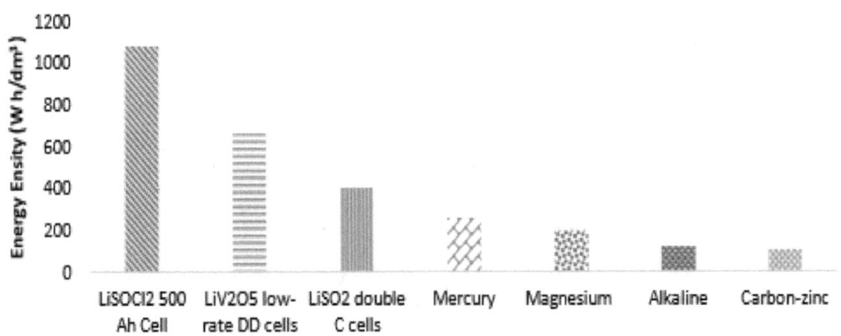

Figure 1. Comparison of energy density of lithium cells and other types of cells.

2.2. *Applications of the Lithium-ion batteries*

The most mature secondary batteries are Lithium-ion (Li-ion) batteries (Massé et al., 2015).

Asahi Kasei initially designed and developed Li-ion batteries in the 1980s, which were later brought to the market by Sony and A&T Battery Corporation in 1991 and 1992. Since then, the popularity of Li-ion batteries has skyrocketed, gaining a significant market share in the mobile electronic devices industry within a few years. Currently, Li-ion battery technology is advancing swiftly to meet the demands of electric vehicles (EVs) and stationary electrical energy storage applications (Bian et al., 2020; Miao et al., 2013).

Due to their interesting features, the Li-Ion are the abundantly used battery types and are the choice for many consumer electronics. They have many advantages including, low rate of loss of charge if not use, and have the highest energy density compared to other batteries. More so, they can come in nearly any size. Apart from being used as batteries for phones, and other hand held gadgets such as smart watches, Lithium ion batteries have other applications, such as in medical devices, in smart electricity grids, in electric vehicles, and other essential, which keep the modern life running . The following are some of the major applications of Lithium-Ion rechargeable batteries.

2.2.1. Emergency power back up or Uninterrupted Power Supply (UPS)

In areas prone to power instability and blackouts, Lithium-ion batteries come in handy since they help to back up power once power is lost. It provides instant power to the systems that were running and helps them to continue running uninterrupted or a chance to shutdown safely until the power is back (Goodenough and Park, 2013). Such a system is crucial in the medical field, and in large communication technology systems. Emergency Power Back Up or Uninterrupted Power Supply (UPS) are used in such emergencies.

2.2.2. Dependable electric and recreational vehicle power in electric vehicles

The durability of Lithium-ion batteries stands out as one of their most significant characteristics., and their efficiency in releasing stored power. This makes them dependable for use in Electric cars. One can use them over a long time, and they lose very little power between uses, hence even better suited for remote locations compared to acid batteries. More so, these Lithium ion batteries are lightweight and of smaller size compared to other power storage technologies thus making them suitable for use in Electric cars (Goodenough and Park, 2013; Qaisar and Alyamani, 2022). Dependable Electric and Recreational Vehicle Power in Electric Vehicles: The durability of Lithium-ion batteries stands out as one of their most significant characteristics, and their efficiency in releasing stored power.

2.2.3. Solar power and wind power storage

The use of renewable energy is gaining traction all over the world. One of the major concerns of renewable technologies such as solar power and wind power is that their production fluctuates. This leads to the need of a power storage system to ensure stable power supply even when the production is low. Lithium-ion batteries come in handy. They are low maintenance cost compared to acid batteries and therefore better-suited (Goodenough and Park, 2013). More so, they charge quickly and release power at a stable pace. Solar Power and Wind Power Storage are used in an increasing number of circumstances.

2.2.4. *Portable power packs (power banks)*

One of the major drawbacks of mobile phones and other portable devices is that they run out of power, especially when visiting or living in remote areas with no access to electricity. Lithium-ion batteries can act as power banks, and use them to store power. Once you need to charge your gadget, you can use the power stored in the Portable Power Packs. Portable Power Packs (Power Banks) are used to charge your gadgets.

2.2.5. *Alarms, surveillance and other security systems*

Lithium batteries are fitted in surveillance, alarms, and other security systems in remote areas since they can be in small, have a longer life span, are rechargeable, and they discharge efficiently. These can be used to ensure that security systems do not go down or stop working even after the grid electricity or other main source of power has stopped working. Alarms, Surveillance and other security systems are used to continue security surveillance.

2.2.6. *Reliabe marine performance*

Boats and small yachts need some source of power when not near land. Such power includes running fridges, and lighting. Diesel generators have been options for sources of power for a long time, but with the increasing consciousness of environmental sustainability, Lithium batteries are becoming an option. Reliabe Marine Performance: Boats and small yachts need some source of power when not near land.

3. The Battery Management System (BMS)

This section of the chapter delves into the importance of BMS utilized in Li-ion battery packs. The BMS plays a critical role in ensuring the safe and reliable operation of Li-ion batteries, as it prevents physical damage, addresses issues related to thermal degradation and cell imbalance, and enables the assessment of various battery conditions such as SoC and SoH (Kim et al., 2017). By effectively detecting temperature, measuring voltage and current, and setting appropriate alarms, an efficient BMS prevents overcharging and over-discharging. Additionally, the BMS plays a vital role in data monitoring and updating, fault detection, and battery voltage equalization, all of which are essential for achieving accurate SoC and SoH readings (Bagade et al., 2022).

The primary objectives of a BMS revolve around ensuring the safety and reliability of the battery pack. All the advanced functionalities integrated into a BMS ultimately serve these core purposes. These features aim to optimize the performance of the battery pack by effectively measuring and monitoring the voltage, current, and temperature of each cell within the pack. Additionally, the BMS

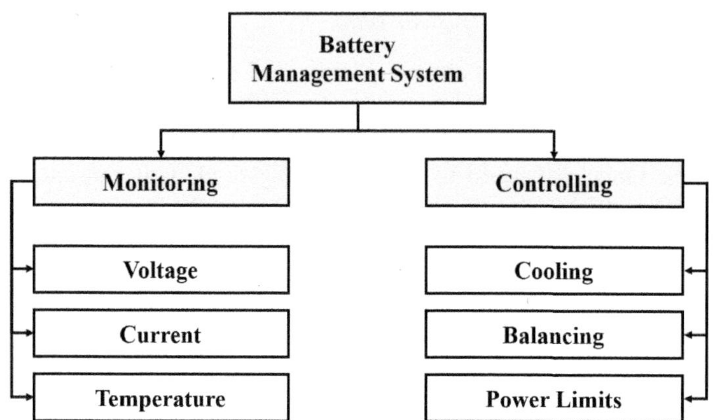

Figure 2. BMS system main functions.

takes charge of regulating cooling systems, balancing cell voltages, and imposing power limits on the battery pack (Pushkar et al., 2022), as well as operates the disconnect breakers for start-up and power-off.

Let us look at the black-box model of a BMS in Figure 2. It can be noted that at a high level and its core, the BMS is taking some inputs and running some algorithms and estimators to produce several outputs to the application system controller. For example, the BMS takes several voltages, such as every cell voltage and pack voltage. It also takes, as an input, The battery pack's current flow, whether it is charging or discharging, and its magnitude are monitored by the Battery Management System (BMS). The BMS utilizes temperature sensors to assess the temperature distribution within the cells. Various algorithms are then employed within the BMS to generate precise estimations. One of the primary outcomes is the estimation of the SoC, commonly referred to as a fuel gauge, which indicates the remaining charge on devices such as phones or electric vehicles. Additionally, the BMS provides an estimation of the SoH. SoH represents the battery pack's capacity relative to its initial state, typically decreasing over time. Accurately predicting the remaining useful life (RUL) of a battery is crucial for intelligent battery health management systems (Ren et al., 2018). The BMS is additionally working on establishing a safe operating envelope (SoE) to determine the allowable current for charging or discharging at any given moment. Furthermore, the BMS sends out fault signals or status indications that the application controller must acknowledge, and thus, specific conditions may activate these signals (Kumar et al., 2015).

In a schematic, a BMS can be used in single or multi-cell Li-ion battery applications. Figure 4 shows a 3S-1P battery pack with three cells connected in series. The BMS is a circuit board that typically resides very close to the cells, monitors the voltage to measure each cell voltage, and monitors the overall battery pack voltage. Besides, the BMS measures the current flowing into or out the pack via a shunt sensor, for example. Finally, the BMS has a master disconnect

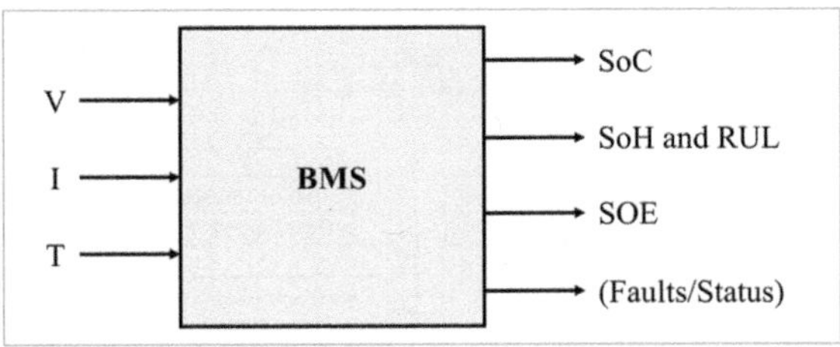

Figure 3. BMS system block diagram.

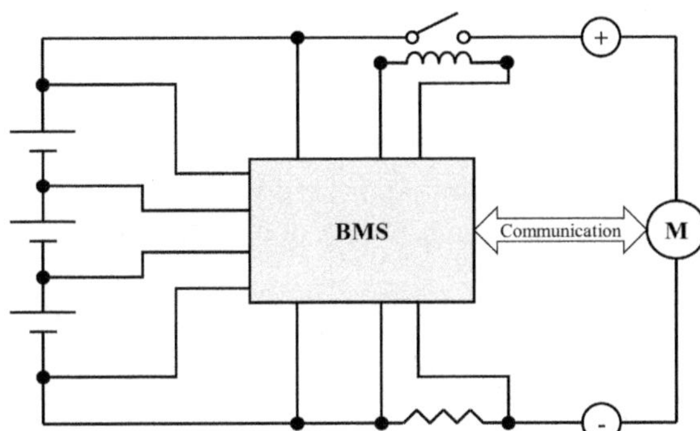

Figure 4. BMS schematic and communication diagram.

that allows it to terminate the battery charge or discharge if it detects that the battery system is entering into an unsafe or undesirable state. Then, external terminals connect the battery pack to the load, i.e., an EV, a solar power inverter, or a DC-DC converter. In a nutshell, the BMS monitors everything going on internally and communicates via an interface to an external controller that uses this information to update its behavior better (Kumar et al., 2015; M. Kokila and Indragandhi, 2020).

3.1. Battery State of Charge (SoC) estimation

In this section of the chapter, the concept of SoC in relation to Li-ion battery packs is explored. SoC estimation has long been a challenging task for all types of energy storage devices. Achieving a highly precise SoC estimation not only provides valuable information regarding the remaining charge or usable energy in the battery, but also offers insights into the battery's reliability. Furthermore, an accurate and efficient SoC estimation plays a crucial role in determining optimal charging and discharging strategies, which greatly impact battery applications. It is important to note that each cell within a battery pack may exhibit different capacities due to factors such as aging, temperature variations, self-discharge, and manufacturing differences (Hannan et al., 2017).

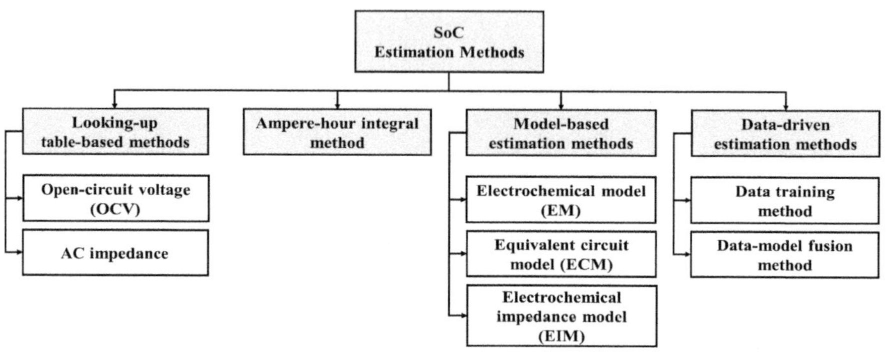

Figure 5. Types of SoC estimation methods (Xiong et al., 2018).

An accurate and effective evaluation of the SoC of a battery plays a crucial role in vehicle energy management and the optimal design of control systems. As a result, various techniques have been proposed to measure the battery's SoC in real-time. To provide a comprehensive analysis of these methods, the authors have categorized them into four distinct groups (Xiong et al., 2018), and this classification is summarized in Figure 5.

The first method used for estimating SoC is the lookup table-based approach, where the SoC of batteries is directly mapped to their external characteristic parameters, such as open-circuit voltage (OCV) and AC impedance. This method is commonly employed in laboratory environments.

The second method is the Ampere-hour integral approach, which relies on accurately measuring the battery's current and knowing its maximum available capacity. By integrating the ampere-hour measurements, the variation in SoC can be accurately calculated. This method is often combined with other techniques, such as model-based methods, to enhance its performance.

Model-based estimation methods form the third category of SoC estimation techniques and can be classified into three subcategories: electrochemical models (EM), equivalent circuit models (ECM), and electrochemical impedance models (EIM). These models leverage the underlying physics and characteristics of the battery to estimate its SoC.

The final method for SoC estimation is the data-driven approach, which constructs a controller using input-output data from the system. Within this category, two types of methods are commonly used. The first is the data training method, which yields highly accurate results but can be sensitive to parameter variations when the training dataset does not adequately cover the current operating conditions. On the other hand, the data-model fusion method combines online data-driven approaches with model-based techniques, ensuring system convergence and stability (Xiong et al., 2018).

3.2. Battery State of Health (SoH) estimation

This section of the chapter will focus on the estimation of the state of health (SoH) in Li-ion battery packs. SoH refers to the battery's ability to store and supply energy based on its initial conditions, taking into account the energy and power requirements of the application. It can be characterized by the state of health related to energy capacity (SoHE) and power capacity (SoHP). SoHE is typically measured by battery capacity, while impedance is used to quantitatively describe SoHP. To ensure the overall system's protection, it is crucial to have a reliable prediction of either SoHE or SoHP using simple methods. By obtaining this information, the optimization can be done on the battery's operating modes, prolong its lifespan, and predict the appropriate timing for battery replacement (Li et al., 2018a).

In (Li et al., 2018a), the battery capacity is determined as the SoH indicator by evaluating the ratio of the current cell capacity to the original cell capacity:

Equation 1 provides a means to determine the SoH percentage of a battery. It involves comparing the current capacity of the cell ($Q_{current}$) to its initial capacity when it was brand new (Q_{fresh}). Initially, when the battery is fresh, its SoH is 100%, representing optimal performance. However, over time, the SoH gradually decreases as the battery degrades.

$$SoH = \frac{Q_{current}(Ah)}{Q_{fresh}(Ah)} \times 100 \tag{1}$$

The specific requirements of an application dictate the end of life (EoL) of the battery. This signifies the point at which the battery can no longer meet the desired performance criteria in terms of capacity or power. For instance, electric and hybrid electric vehicles (EVs and HEVs) necessitate battery replacement when the SoH falls below 80%. However, it's important to note that the direct calculation provided by Equation 1 may not account for all factors influencing battery health is subject to the complete battery charging and discharging cycle. Therefore, this method is not practical in actual application because the batteries, in most cases, are not completely charged and discharged (Li et al., 2018a).

Research on the SoH of batteries has recently attracted significant attention from researchers. Numerous estimation techniques have been developed in this field, each with its own set of advantages and disadvantages concerning estimation accuracy, testing time duration, and implementation feasibility. Figure 6 SoH estimation techniques can be broadly classified into three categories: adaptive models, experimental techniques, and incremental capacity/differential voltage analysis.

The first category, adaptive model-based methods, can be further subdivided into equivalent circuit-based models and electrochemical methods. These models establish a strong physical relationship between the model parameters and the underlying electrochemical processes within the battery cells. However, imple-

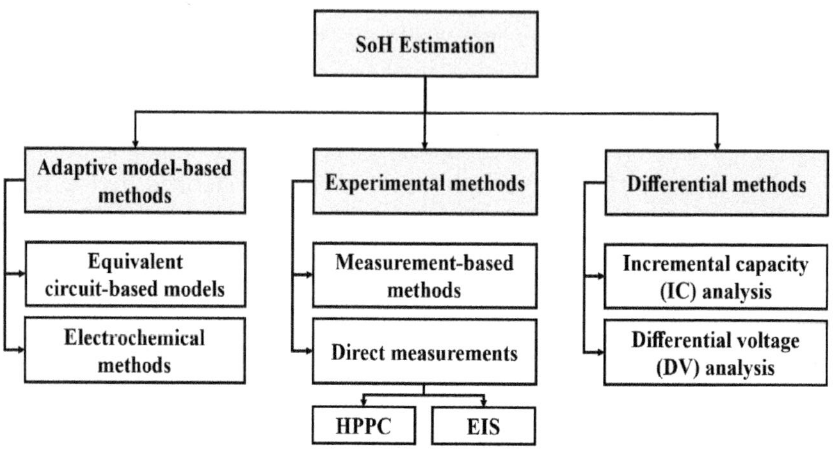

Figure 6. Types of SoH estimation methods (Ali et al., 2023b).

menting model-based techniques in BMS can be challenging due to their computational complexity and the need for large matrix operations.

The second category comprises experimental techniques, which involve measurement-based methods and direct measurements obtained from experiments such as hybrid pulse power characterization (HPPC) and electrochemical impedance spectroscopy (EIS).

The third category encompasses differential methods based on incremental capacity (IC) analysis and differential voltage (DV) analysis. These differential analysis techniques combine the advantages of both experimental and adaptive-model methods. They can not only be used for identifying battery degradation but also enable SoH estimation with low computational effort. However, these methods require static charging/discharging conditions to be fulfilled (Li et al., 2018a).

The authors in (Berecibar et al., 2016) indicate that, Currently, there is no single definitive approach to determining the SoH. The selection of a method should depend on the specific aspect to be estimated and the data that is available. For instance, if there is a large volume of data and a feasible algorithm that can be utilized, a combination of identifying degradation mechanisms and employing robust data techniques would be the most appropriate and satisfactory option. On the other hand, if data is being acquired gradually during the development of tests, it would be more suitable to employ an adaptive model in conjunction with degradation analysis.

3.3. *Battery Remaining Useful Life (RUL) estimation*

The section of the thesis examines recent research on predicting the remaining useful life (RUL) of Li-ion battery packs, which refers to the duration from the present state of the battery to its failure condition, as depicted in Figure 7.

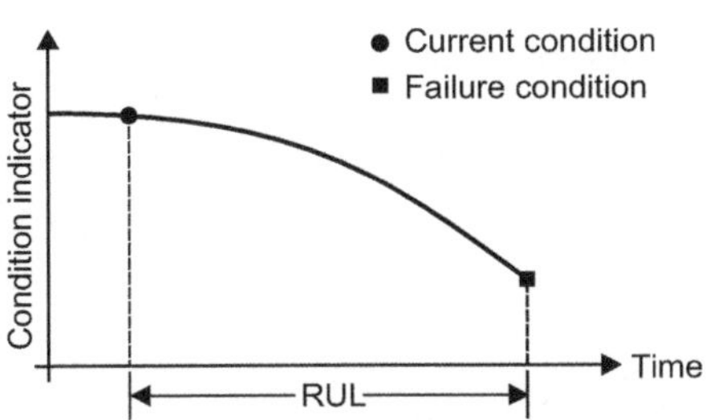

Figure 7. Decaying battery profile.

Many researchers and organizations have intensified their endeavors to enhance the accuracy of estimating the remaining useful life (RUL) of Li-ion batteries. The particle filter (PF) method, also referred to as the sequential Monte Carlo (SMC) method, is a technique that merges Bayesian learning strategies with importance sampling. Qiang et al. (Miao et al., 2013) in 2013, a more advanced variation of the PF method was put forward, which introduced the Unscented Particle Filter (UPF) technique. The UPF approach involves a two-step process. Initially, the proposal distribution is obtained using the Unscented Kalman filter (UKF) algorithm. Subsequently, the PF algorithm is employed to derive the ultimate outcomes. Consequently, a degradation model was designed. The model and algorithms that ran prediction tests recorded an error percentage of less than 5% for actual RUL estimation for Li-ion batteries, which is better than the error percentage recorded using PF algorithms (around 7%) (Miao et al., 2013).

Zhang et al. (Zhang et al., 2018a) in 2018, researchers made significant advancements in the UPF (Uncertainty Propagation Fusion) method for predicting the Remaining Useful Life (RUL) of Li-ion batteries. They introduced a linear optimizing combination resampling technique within the UPF, known as U-LOCR-PF. This innovative approach addressed the limitation of particle diversity found in previous methods.

The study emphasized the importance of the step coefficient K in the LOCR (Local Optimal Combination Resampling) process. The authors highlighted that determining the coefficient K using a fuzzy inference system greatly impacted the performance of LOCR. By optimizing this coefficient, the RUL predictions exhibited enhanced accuracy compared to the previous PF (Particle Filter) and UPF prediction methods.

The results showcased impressive outcomes, with an average root mean squared error (RMSE) of 0.0335 and 0.0201 achieved using 40 and 70 cycles, respectively.

Overall, the researchers' novel approach, U-LOCR-PF, demonstrated improved performance in predicting the RUL of Li-ion batteries, surpassing the limitations of earlier methods.

In numerous studies, scholars have employed regression models as a predictive approach for Li-ion batteries, utilizing statistical methodologies. Long et al. (Long et al., 2013) in 2013, a proposal was made to enhance the autoregressive (AR) model by incorporating the root mean square error (RMSE) as a measure for determining the AR order. The researchers acknowledged the absence of a uniform criterion for determining the AR order and thus utilized an improved particle swarm optimization (PSO) algorithm. By adaptively adjusting the AR model order based on the information extracted from the data, this approach demonstrated accurate prediction of the remaining useful life (RUL) of Li-ion batteries. Additionally, its suitability for onboard applications was established, making it a valuable tool with minimal error margin in RUL prediction. Hu et al. (Hu et al., 2014a) in 2014, a research study was conducted to assess the reliability of rechargeable Li-ion batteries implanted in medical devices. The study employed a developed non-linear kernel regression model, specifically utilizing the k-nearest neighbor (kNN) regression approach. The research focused on analyzing the charging voltage and current curves of the batteries.

By incorporating the kNN regression model, the study was able to capture significant features from the charging curves, as defined by the author. To enhance the accuracy of capacity estimation, the research utilized Particle Swarm Optimization (PSO) to minimize the cross-validation (CV) error. The findings indicated that the proposed model effectively predicted the entire lifespan of the Li-ion battery, as verified by 10 years of cycling data. On average, the model achieved an Root Mean Square Error (RMSE) of 1.08875.

As a data-driven approach based on machine learning, this research demonstrated the efficacy of utilizing the kNN regression model and PSO to accurately estimate the life of Li-ion batteries. Wang et al. (Wang et al., 2013) in 2013, a prognostic model for Li-ion batteries' remaining useful life (RUL) prediction was developed. The model incorporated a relevance vector machine (RVM) algorithm and a capacity degradation model. By utilizing the RVM, the model identified the significant training vector, and three conditional capacity degradation models were created to accurately estimate the predictive values of the relevance vectors. The effectiveness of the model was confirmed through three independent studies, demonstrating satisfactory predictive results. On average, the model achieved an impressive root mean square error (RMSE) of 0.0082.e. Ismail et al. (Li et al., 2014) in 2014, a predictive model utilizing Support Vector Machines (SVM) was proposed to achieve precise Remaining Useful Life (RUL) prediction. The model employed an iterative multi-step approach and considered working temperatures and energy efficiency as input variables to define the training dataset. Extensive experimentation and subsequent analysis of the model demonstrated its ability to accurately identify RUL characteristics of Li-ion batteries using a limited number of parameters.

Patil et al. (Patil et al., 2015) in 2015, a different type of regression model was introduced for the purpose of predicting the real-time remaining useful life (RUL) of Li-ion batteries. This involved the combination of support vector regression (SVR) with a support vector machine (SVM), leveraging their capabilities in both classification and regression tasks. The primary objective of this model was to train an algorithm that could be employed in electric vehicles (EVs) to anticipate when a battery is nearing its end of life (EoL) and notify the driver before complete depletion. To train the model, various cycling data from batteries under different conditions were utilized from the battery dataset. Through conducting case studies, it was determined that this model held promise as a potential onboard tool for estimating RUL in EVs. Zhang et al. (Zhang et al., 2018b) in 2018, a long short-term memory (LSTM) based on a recurrent neural network was introduced for predicting the remaining useful life (RUL) of Li-ion batteries. Experimental data was collected using various Li-ion cells subjected to different currents and temperatures. The LSTM model demonstrated excellent RUL prediction capabilities, regardless of the offline training data.

Miao et al. (Chen et al., 2013) in 2013, a degradation model based on the unified particle filter (PF) approach was developed for the prediction of remaining useful life (RUL) in Li-ion batteries. The model demonstrated enhanced accuracy in RUL prediction compared to the PF method, yielding an error rate of less than 5%.r. Tang et al. (Tang et al., 2014) in 2014, a novel prognostic method utilizing the Wiener process with measurement error (WPME) was developed for predicting the Remaining Useful Life (RUL). Fang et al. (Zheng and Fang, 2015) in 2015, a novel method for predicting Remaining Useful Life (RUL) was proposed, incorporating a non-linear time series prediction model along with the Unscented Kalman Filter (UKF). The battery model state is updated frequently using the UKF and short-term capacity. The results of the proposed model demonstrate superior accuracy and reliability compared to the Extended Kalman Filter (EKF). At 100 cycles, the model achieves a Mean Absolute Percentage Error (MAPE) of 0.1611 and a Root Mean Square Error (RMSE) of 0.01156, validating its effectiveness. Li et al. (Li et al., 2016) in 2016, a state-space model utilizing a spherical cubature particle filter (SCPF) was proposed to analyze the remaining useful life (RUL) of 26 Li-ion batteries. The PF method was outperformed by the proposed model when it came to predicting accuracy. Nevertheless, variations in currents and temperatures could significantly impact the model's precision. Moreover, the authors (Li et al., 2016) the novel prediction method based on Gaussian process mixture (GPM) for Li-ion batteries and Gaussian process regression (GPR) outperforms SVM in terms of reliability and accuracy. The proposed model achieves impressive results, with RMSE values of 0.0158 and 0.0130 at 60 and 80 inspection cycles, respectively. Shen et al. (Shen et al., 2021) in 2020, a novel online approach was introduced for predicting the remaining useful life (RUL) of Li-ion batteries under variable discharge current conditions. The approach utilized a unique two-stage Wiener process model. The authors evaluated this innovative method using two Li-ion batteries and obtained the following performance metrics: a mean absolute error (MAE) of 3 and 2.444, a root mean square error (RMSE) of 3.889 and 3.122, a mean absolute percentage error (MAPE) of 0.0829 and 0.126, and a decision coefficient (R^2) of 0.9751 and 0.9856, respectively, for the first and second battery.

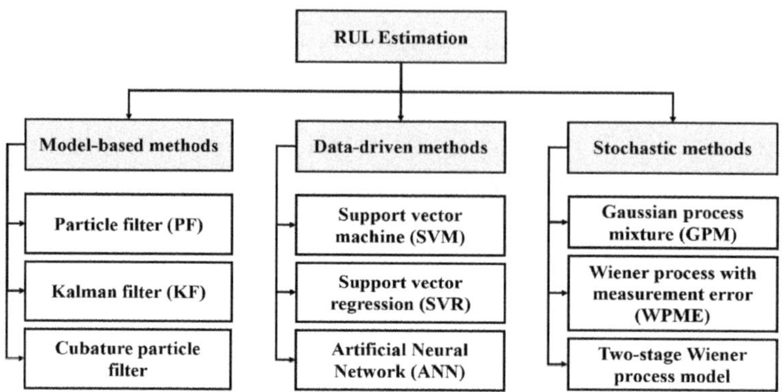

Figure 8. Types of RUL estimation methods.

Figure 8 illustrates the classification of RUL estimation methods for Li-ion batteries into three main categories: model-based, data-driven, and stochastic approaches. The figure provides an overview of the most significant methods discussed in the literature review.

4. AI piloted battery state estimation

Researchers have developed several workflows for battery state estimation, each with its own distinctive approach and methodologies. These workflows are designed to enhance the precision and dependability of battery state estimation by incorporating artificial intelligence techniques. In this discussion, the authors will provide a brief overview of some noteworthy workflows in this domain. The fundamental workflow for battery state estimation generally comprises four steps: data collection and preprocessing, feature engineering, battery model development, and implementation (Catelani et al., 2021; You et al., 2022). Figure 9 (Zhao et al., 2023) provides an overview of this workflow.

4.1. Data collection and preprocessing

The first step is to collect battery state data using a data acquisition system. This can include measurements of voltage, current, temperature, and other relevant parameters. Data preprocessing techniques are then applied to ensure the quality and suitability of the data for further analysis. This may involve tasks such as data cleaning, filtering to remove noise, and regularization to normalize the data (Li et al., 2019a).

4.2. Feature engineering

Feature engineering is centered around the selection and extraction of pertinent features from gathered data. The objective is to diminish data dimensionality,

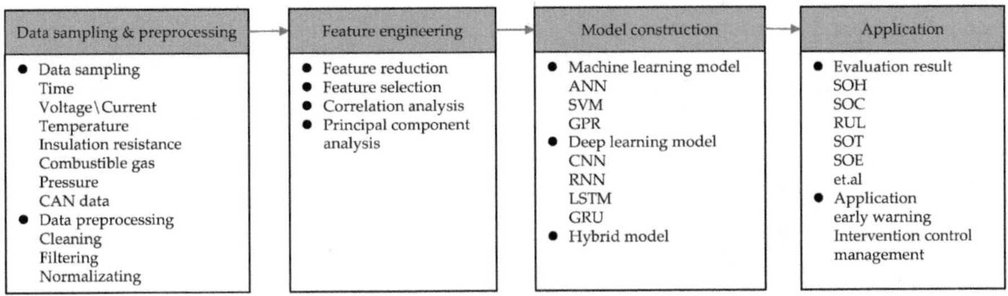

Figure 9. Basic workflow of battery state estimation (Zhao et al., 2023).

capture vital information regarding battery states, and circumvent data redundancy. Approaches such as principal component analysis, correlation coefficient analysis, and cosine similarity analysis can be employed to identify significant features that exhibit a robust correlation with battery states (Jiang et al., 2023).

4.3. Battery model construction

The third step entails creating an estimation model that establishes a mapping relationship between the input data (features) and the output data (battery states). Various modeling approaches can be utilized, including machine learning models, deep learning models, and hybrid models. Machine learning models like support vector regression (SVR), artificial neural networks (ANN), and random forests (RF) are commonly employed. Deep learning models, such as recurrent neural networks (RNN), long short-term memory (LSTM) networks, and convolutional neural networks (CNN), are effective in capturing temporal or spatial dependencies in the data. Hybrid models combine the strengths of both machine learning and deep learning techniques (Cao et al., 2022).

4.4. Application

The final step is to apply the results of battery state estimation for various purposes. This can include identifying abnormal battery states, triggering active interventions or maintenance procedures, and optimizing battery usage. The estimated battery states can be used as inputs for decision-making processes in battery management systems, enabling proactive actions to enhance performance and reliability (Choi et al., 2019).

4.5. Machine learning techniques

Machine learning techniques have garnered considerable interest in battery state estimation because they can learn patterns and correlations from data. These techniques entail training models using past battery data to accurately predict and estimate the state variables.

4.5.1. Support Vector Regression (SVR)

Support Vector Regression (SVR) is a popular machine learning algorithm utilized for estimating the state of batteries. It belongs to the family of support vector machines (SVM) and is specifically designed to handle regression tasks. The primary objective of SVR is to identify the optimal hyperplane that achieves the maximum margin while simultaneously minimizing the discrepancy between the predicted and actual values (Hansen and Wang, 2005).

SVR has proven to be effective in battery state estimation tasks, including the estimation of state of charge SoC and SoH of batteries. By utilizing labeled training data, SVR has the capability to capture intricate connections between battery inputs (such as voltage and current) and outputs (such as SoC), enabling it to deliver precise estimations.

4.5.2. Artificial Neural Networks (ANN)

Artificial Neural Networks (ANN) emulate the structure and functioning of the human brain. They are a widely used machine learning technique for battery state estimation, comprising interconnected layers of artificial neurons (Kang et al., 2014).

Artificial neural network (ANN) models are capable of approximating the intricate nonlinear connections between battery inputs and outputs in battery state estimation. The architecture of ANNs can differ, with feedforward neural networks, recurrent neural networks (RNNs), and convolutional neural networks (CNNs) being common examples. ANNs have showcased remarkable precision when estimating battery state variables like SoC and SoH. These models excel at learning from extensive datasets and comprehending intricate patterns within the data, thereby enabling reliable predictions.

4.5.3. Random Forests (RF)

Random Forests (RF) is a method of ensemble learning that combines multiple decision trees to make predictions. In RF, each tree is constructed using a random subset of the training data. The final prediction is obtained by aggregating the predictions from all the individual trees. RF has demonstrated great potential in battery state estimation tasks because of its ability to handle data with a high number of dimensions and capture complex relationships. It is capable of effectively dealing with noisy data and missing values, making it well-suited for practical battery applications. By leveraging RF, accurate estimations of battery state variables can be achieved, enabling effective management and control of batteries.

4.5.4. Deep Learning Techniques

Deep learning techniques have revolutionized many fields, including battery state estimation. These techniques leverage neural networks with multiple hidden layers to learn complex representations and patterns from data.

4.5.5. Recurrent Neural Networks (RNN)

Recurrent Neural Networks (RNNs) find extensive application in battery state estimation tasks where sequential data, such as time-series battery data, is involved. With their feedback loop mechanism, RNNs can retain information across different time steps, enabling them to grasp the temporal dependencies present in the data. Battery state variables, including SoC and remaining useful life (RUL), have been effectively predicted using RNNs. Their ability to capture long-term dependencies and dynamic patterns makes them a suitable choice for real-time battery state estimation.

4.5.6. Long Short-Term Memory (LSTM) networks

LSTM networks are a specific type of recurrent neural network (RNN) that tackle the issue of the vanishing gradient problem encountered during the training of deep neural networks. LSTMs employ a more intricate structure that incorporates memory cells and gates, enabling them to regulate the information flow. LSTM networks have shown remarkable performance in battery state estimation tasks. They can effectively capture long-term dependencies and handle the challenges of noisy and irregularly sampled battery data. LSTMs have been used for accurate SoC estimation and RUL prediction (Chinomona et al., 2020).

4.5.7. Convolutional Neural Networks (CNN)

Convolutional Neural Networks (CNNs) find their primary application in image processing, but they have also been utilized for battery state estimation.CNNs excel in capturing spatial relationships in the data and have achieved significant success in image-based battery state estimation, such as estimating SoH based on battery images (Jiang et al., 2023).

By applying convolutional operations to battery data, CNNs can learn spatial features and patterns that contribute to accurate state estimation. CNNs have the advantage of being able to handle multi-dimensional data, making them suitable for tasks involving multi-sensor battery data (Tao et al., 2020).

4.6. Review of related studies

Neural Networks (NN) demonstrate excellent capabilities in creating non-linear maps to represent complex nonlinear battery models. For instance, in (Dang et al., 2016). A proposed method for estimating SoC involves the fusion of two neural networks (NN) within a battery model. The first NN, a linear model, identifies electrochemical parameters of the battery's electrochemical model. The second NN, a backpropagation NN (BPNN), captures the correlation between open circuit voltage (OCV) and SoC. In (Sun et al., 2016) an uncertainty quantification algorithm utilizing a Radial Basis Function Neural Network (RBFNN) was developed to estimate SoC for battery packs comprising multiple cells. The study in (Tong et al., 2016) established a load classifying NN model with improved post-processing to suppress overfitting. A deep feed-forward NN-based method directly mapped battery measurements to SoC (Chemali et al., 2018). Neural networks (NN) have emerged as a popular method in various recent studies.

In (Li et al., 2020), Particle Swarm Optimization (PSO) was utilized to fine-tune the penalty factor and kernel width in Support Vector Regression (SVR) for SoC estimation. These particular parameters have a substantial influence on the performance of SVR. The optimized model demonstrated superior results compared to traditional SVR approaches, achieving an average estimation error of 1.5%. In (Hu et al., 2014b) the authors have employed a Gaussian kernel and fine-tuned cost parameters using a double search algorithm. It resulted in an improved performance with a maximum Mean Square Error (MSE) of 2.23% compared to different variations of Feedforward Neural Networks (FNNs) trained on various datasets. In (Zhang and Wang, 2018), the authors developed an online SVR model to enhance the BMS by providing real-time updates on the battery cell's state. This led to a decrease in the error range and Root Mean Squared Error (RMSE) when compared to Backpropagation Neural Networks (BPNN). On the other hand, in (Xuan et al., 2020) observed that Support Vector Machines (SVMs) exhibit higher accuracy for classification tasks than for regression.

CNNs, initially introduced by (Lecun et al., 1998) as LeNet-5Deep feedforward architectures, such as LeNet-5, fall under the category of convolutional neural networks (CNNs). LeNet-5 consists of two convolutional layers, two pooling layers, two fully connected layers, and an output layer. CNNs have gained immense popularity in image and video processing applications due to their remarkable ability to effectively capture data patterns within a limited period. It is made possible by their deep architecture, which allows them to extract temporal information from training data, setting them apart from feedforward neural networks (FNNs) based on multilayer perceptron (MLP) models.). In (Kuang and Xu, 2018) the authors were pioneers in the application of LeNet-5 CNNs to machine learning tasks involving one-dimensional data. A typical CNN architecture comprises an input layer and one or more convolutional layers with pooling, as shown in Figure 10. The configuration of the output layers varies depending on the specific application. In the case of regression problems, a flattening layer followed by one or more dense layers is commonly used to determine the output. Each convolutional layer contains multiple filters or kernels (referred to as "f") whose size is determined by the input dimensionality.

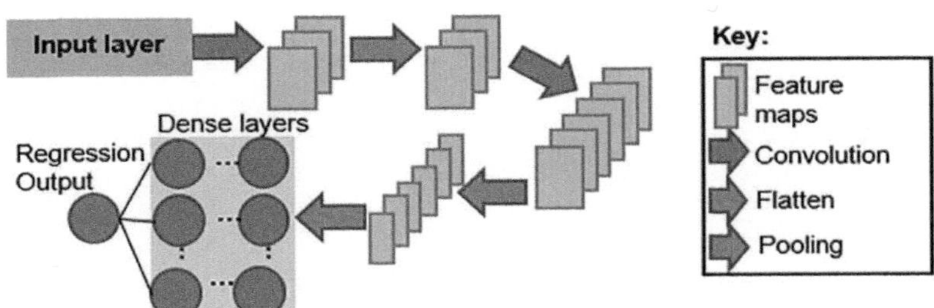

Figure 10. CNN architecture.

Several models, such as linear regression, linear support vector machine (SVM), k-nearest neighbors (kNN), random forest (RF), and LightGBM (LGBM), were assessed. Linear SVM seeks to find a linear hyperplane that effectively fits all data points across all dimensions. kNN makes predictions based on the labels of its k nearest neighbors. RF utilizes multiple decision tree regressors, which are susceptible to overfitting, and implements bagging with majority voting to reduce variance. LGBM employs multiple tree regressors trained on the entire dataset to address underfitting and minimize bias by focusing on error reduction (Granado et al., 2022).

Linear regression did not undergo any optimization, while other models were subjected to coarse optimization using grid search. The hyperparameters that were tuned include regularization and ϵ for SVM, the number of neighbors (k) for kNN, and the number of trees and maximum features for RF. As LGBM requires tuning a larger number of hyperparameters, a randomized search was conducted, testing 50 combinations of hyperparameters such as leaf number, minimum child samples, minimum child weight, subsample ratio, subsample ratio of columns for each tree, and α (L1) and λ (L2) regularization parameters (Probst et al., 2019).

The findings from the model selection are illustrated in Figure 11, where (a) to (e) showcase a comparison between the actual SoH values of the test set and the corresponding predicted values. Figure 11f provides a comprehensive overview of the results, presenting scores and fit times. All models exhibited excellent performance on the training set, with coefficients of determination (train R2) exceeding 0.992, mean absolute errors (train MAE) below 0.005 (equivalent to 0.5%), and root mean squared errors (train RMSE) below 0.007 (equivalent to 0.7%). Notably, the performance scores of the models consistently surpassed those reported in previous studies, as affirmed in the literature review (Chandran et al., 2021).

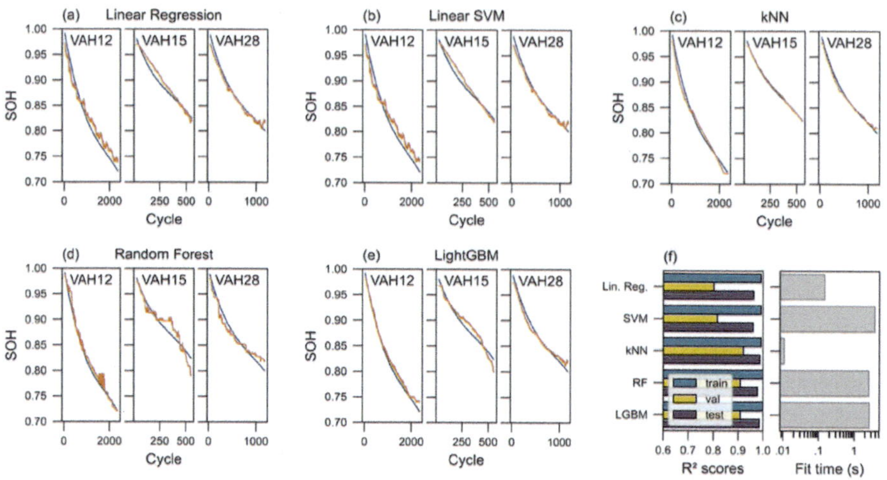

Figure 11. Displays the true (blue) and predicted (orange) SoH values as a function of cycle for the three cells in the test set, utilizing different models. Additionally, the graph displays the R2 scores for training, validation, and testing, as well as the mean fit times for all models. It is important to note that the SoH predictions presented here are based on the thirty most recent observations, which are included in the rolling window. For a visual reference to the color scheme, please refer to the online version of this article.

5. Discussion

The AI assistive state estimation of rechargeable battery cells has gained a significant attention due to its crucial role in optimizing the performance, efficiency, and lifespan of the intended battery packs (Faraji-Niri et al., 2023). Among different battery technologies, the Li-Ion batteries are widely used in various applications such as electric vehicles, renewable energy storage, satellites, mobile phones, and emergency power backup systems. It is due to their ever-wanted features such as the high power density, lighter weight, smaller size, and a higher count of charge-discharge cycles. The Li-Ion batteries are relatively costly however an effective usage can render a longer battery life and a viable solution. In this context, an Accurate estimation of the battery cell's SoC, capacity, SoH, and RUL are vital for effective battery management and usage (Faraji-Niri et al., 2023).

The BMS play a critical role in monitoring and controlling the operation of Li-Ion batteries (Sun et al., 2021). A BMS consists of various components such as sensors, control algorithms, and communication interfaces to ensure the safety, reliability, and longevity of the battery (Sun et al., 2021). One of the key functions of a BMS is to estimate the SoC, which represents the available energy remaining in a battery (Wang et al., 2021).

AI-driven state estimation techniques have shown great potential in enhancing the performance and reliability of Li-Ion batteries. Accurate estimation of SoC, capacity, RUL, and SoH can significantly improve battery management, optimize charging and discharging strategies, and extend the battery's lifespan. Furthermore, these techniques enable the implementation of predictive maintenance strategies, allowing for proactive replacement of aging cells before failure occurs (Guo and Ma, 2023).

The advantages of AI technology are numerous. While avoiding a difficult battery modelling procedure, these approaches may be utilized to evaluate the status of battery cells, extract online aging data from measurements, and connect it to battery performance characteristics. In comparison to existing industrial standards, the accuracy of battery health predictions, made using AI approaches, is substantially greater. Additionally, performance data is being generated by the AI in order to train it how to create better batteries. Artificial intelligence and extensive experimental data may be used to uncover the secret to correctly forecasting the usable life of Li-Ion battery cells before their capacities start to decline. However, due to significant alterations in battery properties throughout the course of a battery's lifespan brought on by aging and unique nonlinear behaviour, estimating the state of the battery cell is a difficult process. As a result, the results of AI-based models may be inaccurate, particularly when there is a lack of training data.

The future of current energy generating, and distribution networks is the Internet of Energy (IoE). It will be built on modern AI, smart meters, and information and communication technology (ICT). Data gathering with the right level of granularity is crucial in this infrastructure (Qaisar and Alsharif, 2020; Qaisar et al., 2019a). The massive deployment of smart meters can lead towards an exponential raise in the collected data (Qaisar and Alsharif, 2020). It can lead towards data management and transmission challenges. In this context, the event-driven data acquisition and processing can render efficient realizations in terms of compression and computational plus transmission efficiencies (Qaisar and Aljefri, 2021; Qaisar et al., 2019b; Sabo et al., 2018). Additionally, the optimization tools can result in effective data dimensionality reduction without losing the pertinent information (Khan et al., 2022; Mian Qaisar et al., 2023; Qaisar et al., 2022). Future research can look in the viability of incorporating these technologies in battery management systems.

Our research, which primarily delves into Li-ion battery state estimation using artificial intelligence techniques, is marked by several noteworthy limitations. Firstly, its narrow scope, centering solely on Li-ion batteries, neglects the exploration of other rechargeable battery types and alternative state estimation methods, potentially limiting the broader applicability of our findings. Secondly, the effectiveness of artificial intelligence algorithms hinges on data availability, yet we grapple with challenges in obtaining substantial real-world battery data, raising concerns about the generalizability of our conclusions. Thirdly, certain AI-driven techniques, especially model-based approaches, come with significant computational complexity, posing practical challenges for their real-time implementation in battery management systems. Furthermore, our research, while rich in theoretical discussions and references to existing studies, lacks empirical validation or experimental testing, leaving uncertainties regarding the real-world performance of the discussed methods. Lastly, our study's focus on specific applications, such as electric vehicles, renewable energy, and portable electronics, introduces a limitation in terms of generalizability, given the considerable variations in battery characteristics and requirements across different industries. These limitations underscore the need for continued exploration and improvement in battery state estimation research.

6. Conclusion

This chapter has provided an in-depth exploration of artificial intelligence assistive Li-Ion battery cell's state estimation approaches. The chapter began with the purpose of battery state estimation. Onward, its importance is highlighted in the battery management systems while considering several key applications. Afterward, the advances in battery state estimation due to the artificial intelligence are described. The use of artificial intelligence driven battery management approaches are reviewed. Different rechargeable battery technologies are presented. It is mentioned that due to pertinent benefits the Li-Ion batteries are the most frequently used ones and their several applications are also presented. Various key methods sued for the battery state of charge, state of health, capacity and remaining useful life estimation are summarized. Then a focused discussion on the AI piloted battery state estimation is made. Its key steps such as the data collection, preprocessing, feature engineering, classification/regression, and intended application are described. The key machine/ensemble/deep learning techniques, used in battery state estimation are outlined. A review of related studies is also made while describing the advantages and limitations of the artificial intelligence based rechargeable battery cell's state estimation. As the field continues to advance, future directions may involve exploring advanced and contemporary energy storage technologies like supercapacitors, optimizing the integration of energy storage with renewable sources, and developing the Internet of Energy functionalities to further enhance the efficiency, reliability, and sustainability of modern grid systems.

Abbreviations

AI	Artificial Intelligence
ANN	Artificial Neural Network
BMS	Battery Management System
CV	Cross Validation
EIM	Electrochemical Impedance model
EKF	Extended Kalman Filter
EOL	End of Life
EVs	Electric Vehicle
FNN	Feedforward Neural Network
GAN	Generative Adversarial Network
GCD	Greatest Common Divisor
LCM	Least Common Multiple
Li-ion	Lithium Ion
LSTM	Long Short-Term Memory
MAPE	Mean Absolute Percentage Error
OCV	Open Circuit Voltage
PSO	Particle Swarm Optimization
RMSE	Root Mean Squared Error
RUL	Remaining Useful Life
SoC	State of Charge
SoH	State of Health
SoP	State of Power
SVM	Support Vector Machine
UKF	Unscented Kalman Filter

References

Aldosary, A., Ali, Z. M., Alhaider, M. M., Ghahremani, M., Dadfar, S. and Suzuki, K. 2021. A modified shuffled frog algorithm to improve mppt controller in pv system with storage batteries under variable atmospheric conditions. Control Engineering Practice, 112: 104831. ISSN 0967-0661. URL https://www.sciencedirect.com/science/article/pii/S0967066121001088.

Ali, Z. M., Calasan, M., Aleem, S. H. E. A., Jurado, F. and Gandoman, F. H. 2023a. Applications of energy storage systems in enhancing energy management and access in microgrids: A review. Energies, 16(16). ISSN 1996-1073. URL https://www.mdpi.com/1996-1073/16/16/5930.

Ali, Z. M., Calasan, M., Gandoman, F. H., Jurado, F. and Abdel Aleem, S. H. 2023b. Review of batteries reliability in electric vehicle and e-mobility applications. Ain Shams Engineering Journal, p. 102442. ISSN 2090-4479. URL https://www.sciencedirect.com/science/article/pii/S2090447923003313.

Alirezazadeh, A., Rashidinejad, M., Afzali, P. and Bakhshai, A. R. 2021. A new flexible and resilient model for a smart grid considering joint power and reserve scheduling, vehicle-to-grid and demand response. Sustainable Energy Technologies and Assessments, 43: 100926. URL https://api.semanticscholar.org/CorpusID:229412971.

Álvarez Antón, J. C., García Nieto, P. J., Blanco Viejo, C. and Vilán Vilán, J. A. 2013. Support vector machines used to estimate the battery state of charge. IEEE Transactions on Power Electronics, 28(12): 5919–5926.

Bachman, J. C., Muy, S., Grimaud, A., Chang, H. -H., Pour, N., Lux, S. F., Paschos, O., Maglia, F., Lupart, S., Lamp, P., Giordano, L. and Shao-Horn, Y. 2016. Inorganic solid-state electrolytes for lithium batteries: Mechanisms and properties governing ion conduction. Chemical Reviews, 116(1): 140–162. URL https://doi.org/10.1021/acs.chemrev.5b00563. PMID: 26713396.

Bagade, A. C., Gambhir, A. V. and Mandhana, A. 2022. State of charge estimation of li-ion battery using extended kalman filter and combined battery model. Technical Report, SAE Technical Paper.

Berecibar, M., Gandiaga, I., Villarreal, I., Omar, N., Van Mierlo, J. and P. Van den Bossche, P. 2016. Critical review of state of health estimation methods of li-ion batteries for real applications. Renewable and Sustainable Energy Reviews, 56: 572–587. ISSN 1364-0321. URL https://www.sciencedirect.com/science/article/pii/S1364032115013076.

Bian, X., Liu, L., Yan, J., Zou, Z. and Zhao, R. 2020. An open circuit voltage-based model for state-of-health estimation of lithium-ion batteries: Model development and validation. Journal of Power Sources, 448: 227401. ISSN 0378-7753. URL https://www.sciencedirect.com/science/article/pii/S0378775319313941.

Bruce, P., Freunberger, S., Hardwick, L. and Tarascon, J. -M. 2011. Li–o2 and li–s batteries with highenergy storage. Nature Materials, 11: 19–29, 12.

Buchmann, I. 2001. Batteries in a Portable World: A Handbook on Rechargeable Batteries for Nonengineers. 2nd ed., Jul 2001.

Cao, R., Cheng, H., Jia, X., Gao, X., Zhang, Z., Wang, M., Li, S., Zhang, C., Ma, B., Liu, X. et al. 2022. Non-invasive characteristic curve analysis of lithium-ion batteries enabling degradation analysis and data-driven model construction: A review. Automotive Innovation, 5(2): 146–163.

Catelani, M., Ciani, L., Fantacci, R., Patrizi, G. and Picano, B. 2021. Remaining useful life estimation for prognostics of lithium-ion batteries based on recurrent neural network. IEEE Transactions on Instrumentation and Measurement, 70: 1–11.

Chandran, V., Patil, C. K., Karthick, A., Ganeshaperumal, D., Rahim, R. and Ghosh, A. 2021. State of charge estimation of lithium-ion battery for electric vehicles using machine learning algorithms. World Electric Vehicle Journal, 12(1). ISSN 2032-6653. URL https://www.mdpi.com/2032-6653/12/1/38.

Chemali, E., Kollmeyer, P. J., Preindl, M. and Emadi, A. 2018. State-of-charge estimation of li-ion batteries using deep neural networks: A machine learning approach. Journal of Power Sources, 400: 242–255. ISSN 0378-7753. URL https://www.sciencedirect.com/science/article/pii/S0378775318307080.

Chen, T., Jin, Y., Lv, H., Yang, A., Liu, M., Chen, B., Xie, Y. and Chen, Q. 2020. Applications of lithium ion batteries in grid-scale energy storage systems. Transactions of Tianjin University, 26: 208–217. URL https://api.semanticscholar.org/CorpusID:214113614.

Chen, Y., Miao, Q., Zheng, B., Wu, S. and Pecht, M. 2013. Quantitative analysis of lithium-ion battery capacity prediction via adaptive bathtub-shaped function. Energies, 6(6): 3082–3096. ISSN 1996-1073. URL https://www.mdpi.com/1996-1073/6/6/3082.

Chinomona, B., Chung, C., Chang, L. -K., Su, W. -C. and Tsai, M. -C. 2020. Long short-term memory approach to estimate battery remaining useful life using partial data. IEEE Access, 8: 165419–165431.

Choi, Y., Ryu, S., Park, K. and Kim, H. 2019. Machine learning-based lithium-ion battery capacity estimation exploiting multi-channel charging profiles. IEEE Access, 7: 75143–75152.

Dang, X., Yan, L., Xu, K., Wu, X., Jiang, H. and Sun, H. 2016. Open-circuit voltage-based state of charge estimation of lithium-ion battery using dual neural network fusion battery model. Electrochimica Acta, 188: 356–366. ISSN 0013-4686. URL https://www.sciencedirect.com/science/article/pii/S0013468615309257.

Fan, X., Zhang, W., Zhang, C., Chen, A. and An, F. 2022. Soc estimation of li-ion battery using convolutional neural network with u-net architecture. Energy, 256: 124612. ISSN 0360-5442. URL https://www.sciencedirect.com/science/article/pii/S0360544222015158.

Faraji-Niri, M., Rashid, M., Sansom, J., Sheikh, M., Widanage, D. and Marco, J. 2023. Accelerated state of health estimation of second life lithium-ion batteries via electrochemical impedance spectroscopy tests and machine learning techniques. Journal of Energy Storage, 58: 106295. ISSN 2352-152X. URL https://www.sciencedirect.com/science/article/pii/S2352152X22022848.

Galiounas, E., Tranter, T. G., Owen, R. E., Robinson, J. B., Shearing, P. R. and Brett, D. J. 2022. Battery state-of-charge estimation using machine learning analysis of ultrasonic signatures. Energy and AI, 10: 100188. ISSN 2666-5468. URL https://www.sciencedirect.com/science/article/pii/S2666546822000374.

Gandoman, F. H., Ahmed, E. M., Ali, Z. M., Berecibar, M., Zobaa, A. F. and Abdel Aleem, S. H. E. 2021. Reliability evaluation of lithium-ion batteries for e-mobility applications from practical and technical perspectives: A case study. Sustainability, 13(21). ISSN 2071-1050. URL https://www.mdpi.com/2071-1050/13/21/11688.

Gandoman, F. H., El-Shahat, A., Alaas, Z. M., Ali, Z. M., Berecibar, M. and Abdel Aleem, S. H. E. 2022. Understanding voltage behavior of lithium-ion batteries in electric vehicles applications. Batteries, 8(10). ISSN 2313-0105. URL https://www.mdpi.com/2313-0105/8/10/130.

Goodenough, J. B. and Park, K. -S. 2013. The li-ion rechargeable battery: A perspective. Journal of the American Chemical Society, 135(4): 1167–1176. URL https://doi.org/10.1021/ja3091438. PMID: 23294028.

Granado, L., Ben-Marzouk, M., Solano Saenz, E., Boukal, Y. and Jugé, S. 2022. Machine learning predictions of lithium-ion battery state-of-health for evtol applications. Journal of Power Sources, 548: 232051. ISSN 0378-7753. URL https://www.sciencedirect.com/science/article/pii/S037877532201028X.

Gungor, V. C., Sahin, D., Kocak, T., Ergut, S., Buccella, C., Cecati, C. and Hancke, G. P. 2011. Smart grid technologies: Communication technologies and standards. IEEE Transactions on Industrial Informatics, 7(4): 529–539.

Guo, S. and Ma, L. 2023. A comparative study of different deep learning algorithms for lithium-ion batteries on state-of-charge estimation. Energy, 263: 125872. ISSN 0360-5442. URL https://www.sciencedirect.com/science/article/pii/S036054422202758X.

Hannan, M., Lipu, M., Hussain, A. and Mohamed, A. 2017. A review of lithium-ion battery state of charge estimation and management system in electric vehicle applications: Challenges and recommendations. Renewable and Sustainable Energy Reviews, 78: 834–854. ISSN 1364-0321. URL https://www.sciencedirect.com/science/article/pii/S1364032117306275.

Hansen, T. and Wang, C. -J. 2005. Support vector based battery state of charge estimator. Journal of Power Sources, 141(2): 351–358. ISSN 0378-7753.URL https://www.sciencedirect.com/science/article/pii/S0378775304010626.

Hassan, M. 2023. A stochastic approach for environmental operation in microgrids considering parameters uncertainty. International Journal of Advanced Engineering and Business Sciences, URL https://api.semanticscholar.org/CorpusID:256916810.

Hossain Lipu, M., Ansari, S., Miah, M. S., Meraj, S. T., Hasan, K., Shihavuddin, A., Hannan, M., Muttaqi, K. M. and Hussain, A. 2022. Deep learning enabled state of charge, state of health and remaining useful life estimation for smart battery management system: Methods, implementations, issues and prospects. Journal of Energy Storage, 55: 105752. ISSN 2352-152X. URL https://www.sciencedirect.com/science/article/pii/S2352152X22017406.

Hu, C., Jain, G., Zhang, P., Schmidt, C., Gomadam, P. and Gorka, T. 2014a. Data-driven method based on particle swarm optimization and k-nearest neighbor regression for estimating capacity of lithium-ion battery. Applied Energy, 129: 49–55. ISSN 0306-2619. URL https://www.sciencedirect.com/science/article/pii/S0306261914004401.

Hu, J., Hu, J., Lin, H., Li, X., Jiang, C., Qiu, X. and Li, W. 2014b. State-of-charge estimation for battery management system using optimized support vector machine for regression. Journal of Power Sources, 269: 682–693. ISSN 0378-7753. URL https://www.sciencedirect.com/science/article/pii/S0378775314010593.

Hu, X., Feng, F., Liu, K., Zhang, L., Xie, J. and Liu, B. 2019. State estimation for advanced battery management: Key challenges and future trends. Renewable and Sustainable Energy Reviews, 114: 109334. ISSN 1364-0321. URL https://www.sciencedirect.com/science/article/pii/S1364032119305428.

Iqbal, H., Sarwar, S., Kirli, D., Shek, J. K. H. and Kiprakis, A. 2023. A survey of second-life batteries based on techno-economic perspective and applications-based analysis. Carbon Neutrality, 2. URL https://api.semanticscholar.org/ CorpusID:257902829.

Jiang, Y., Chen, Y., Yang, F. and Peng, W. 2023. State of health estimation of lithium-ion battery with automatic feature extraction and self-attention learning mechanism. Journal of Power Sources, 556: 232466. ISSN 0378-7753. URL https://www.sciencedirect.com/science/article/pii/S0378775322014434.

Kang, L., Zhao, X. and Ma, J. 2014. A new neural network model for the state-of-charge estimation in the battery degradation process. Applied Energy, 121: 20–27. ISSN 0306-2619. URL https://www.sciencedirect.com/science/article/pii/ S03062619914000956.

Khan, H., Nizami, I. F., Qaisar, S. M., Waqar, A., Krichen, M. and Almaktoom, A. T. 2022. Analyzing optimal battery sizing in microgrids based on the feature selection and machine learning approaches. Energies, 15(21). ISSN 1996-1073. URL https://www.mdpi.com/1996-1073/15/21/7865.

Kim, Y., Yun, S. and Lee, J. 2017. SoC estimation and bms design of li-ion battery pack for driving. In 2017 14th International Conference on Ubiquitous Robots and Ambient Intelligence (URAI), pp. 216–218.

Kokila, P. M. M. and Indragandhi, V. 2020. Design and development of battery management system (bms) using hybrid multilevel converter. International Journal of Ambient Energy, 41(7): 729–737. URL https://doi.org/10.1080/0143 0750.2018.1492440.

Konatowski, S., Kaniewski, P. and Matuszewski, J. 2016. Comparison of estimation accuracy of ekf, ukf and pf filters. 23(1): Dec 2016.

Krichen, M., Basheer, Y., Qaisar, S. M. and Waqar, A. 2023. A survey on energy storage: Techniques and challenges. Energies, 16(5). ISSN 1996-1073. URL https://www.mdpi.com/1996-1073/16/5/2271.

Krishnamoorthy, U., Gandhi Ayyavu, P., Panchal, H., Shanmugam, D., Balasubramani, S., Al-rubaie, A. J., Al-khaykan, A., Oza, A. D., Hembrom, S., Patel, T., Vizureanu, P. and Burduhos-Nergis, D. -P. 2023. Efficient battery models for performance studies-lithium ion and nickel metal hydride battery. Batteries, 9(1). ISSN 2313-0105. URL https://www.mdpi.com/2313-0105/9/1/52.

Kuang, D. and Xu, B. 2018. Predicting kinetic triplets using a 1d convolutional neural network. Thermochimica Acta, 669: 8–15. ISSN 0040-6031. URL https:// www.sciencedirect.com/science/article/pii/S0040603118304325.

Kumar, B., Khare, N. and Chaturvedi, P. K. 2015. Advanced battery management system using matlab/simulink. In 2015 IEEE International Telecommunications Energy Conference (INTELEC), pp. 1–6.

Kundur, P. and Malik, O. 2022. Power system stability and control. The EPRI power system engineering series. McGraw-Hill Education. ISBN 9781260473544. URL https://books.google.com.sa/books?id=1-WazgEACAAJ.

Kühnelt, H., Beutl, A., Mastropierro, F., Laurin, F., Willrodt, S., Bismarck, A., Guida, M. and Romano, F. 2022. Structural batteries for aeronautic applications—state of the art, research gaps and technology development needs. Aerospace, 9(1). ISSN 2226-4310. URL https://www.mdpi.com/2226-4310/9/1/7.

Lecun, Y., Bottou, L., Bengio, Y. and Haffner, P. 1998. Gradient-based learning applied to document recognition. Proceedings of the IEEE, 86(11): 2278–2324.

Li, D. Z., Wang, W. and Ismail, F. 2014. A mutated particle filter technique for system state estimation and battery life prediction. IEEE Transactions on Instrumentation and Measurement, 63(8): 2034–2043.

Li, L., Wang, P., Chao, K. -H., Zhou, Y. and Xie, Y. 2016. Remaining useful life prediction for lithium-ion batteries based on gaussian processes mixture. PLOS ONE, 11(9): 1–13. URL https://doi.org/10.1371/journal.pone.0163004.

Li, R., Xu, S., Li, S., Zhou, Y., Zhou, K., Liu, X. and Yao, J. 2020. State of charge prediction algorithm of lithium-ion battery based on pso-svr cross validation. IEEE Access, 8: 10234–10242.

Li, S., He, H. and Li, J. 2019a. Big data driven lithium-ion battery modeling method based on sdae-elm algorithm and data pre-processing technology. Applied Energy, 242: 1259–1273. ISSN 0306-2619. URL https://www.sciencedirect.com/science/article/pii/S0306261919305495.

Li, X., Zhang, L., Wang, Z. and Dong, P. 2019b. Remaining useful life prediction for lithium-ion batteries based on a hybrid model combining the long short-term memory and elman neural networks. 21, Feb 2019.

Li, Y., Abdel-Monem, M., Gopalakrishnan, R., Berecibar, M., Nanini-Maury, E., Omar, N., van den Bossche, P. and Van Mierlo, J. 2018a. A quick on-line state of health estimation method for li-ion battery with incremental capacity curves processed by gaussian filter. Journal of Power Sources, 373: 40–53. ISSN 0378-7753. URL https://www.sciencedirect.com/science/article/pii/S0378775317314532.

Li, Y., Zou, C., Berecibar, M., Nanini-Maury, E., Chan, J. C. -W., van den Bossche, P., Van Mierlo, J. and Omar, N. 2018b. Random forest regression for online capacity estimation of lithium-ion batteries. Applied Energy, 232: 197–210. ISSN 0306-2619. URL https://www.sciencedirect.com/science/article/pii/S0306261918315010.

Ling, C. 2022. A review of the recent progress in battery informatics. 8(1), Feb 2022.

Liu, W., Placke, T. and Chau, K. 2022. Overview of batteries and battery management for electric vehicles. Energy Reports, 8: 4058–4084. ISSN 2352-4847. URL https://www.sciencedirect.com/science/article/pii/S2352484722005716.

Long, B., Xian, W., Jiang, L. and Liu, Z. 2013. An improved autoregressive model by particle swarm optimization for prognostics of lithium-ion batteries. Microelectronics Reliability, 53(6): 821–831. ISSN 0026-2714. URL https://www.sciencedirect.com/science/article/pii/S002627141300022X.

Lyu, Z., Lim, G. J., Koh, J. J., Li, Y., Ma, Y., Ding, J., Wang, J., Hu, Z., Wang, J., Chen, W. and Chen, Y. 2021. Design and manufacture of 3d-printed batteries. Joule, 5(1): 89–114. ISSN 2542-4351. URL https://www.sciencedirect.com/science/article/pii/S2542435120305183.

Manoharan, A., Begam, K., Aparow, V. R. and Sooriamoorthy. D. 2022. Artificial neural networks, gradient boosting and support vector machines for electric vehicle battery state estimation: A review. Journal of Energy Storage, 55: 105384. ISSN 2352-152X. URL https://www.sciencedirect.com/science/article/pii/S2352152X22013780.

Mansuroglu, A., Gencten, M., Arvas, M. B., Sahin, M. and Sahin, Y. 2023. Investigation the effects of chlorine doped graphene oxide as an electrolyte additive for gel type valve regulated lead acid batteries. Journal of Energy Storage, 64: 107224. ISSN 2352-152X. URL https://www.sciencedirect.com/science/article/pii/S2352152X23006217.

Massé, R. C., Uchaker, E., Cao, G., Cao, G. and Cao, G. 2015. Beyond li-ion: Electrode materials for sodium- and magnesium-ion batteries. 58(9), Sep 2015.

Mian Qaisar, S., AbdelGawad, A. E. E. and Srinivasan, K. 2021. Event-driven acquisition and machine-learning-based efficient prediction of the li-ion battery capacity. SN Comput. Sci., 3(1): oct 2021. URL https://doi.org/10.1007/s42979-021-00905-0.

Mian Qaisar, S., Khan, S. I., Srinivasan, K. and Krichen, M. 2023. Arrhythmia classification using multirate processing metaheuristic optimization and variational mode decomposition. Journal of King Saud University - Computer and Information Sciences, 35(1): 26–37. ISSN 1319-1578. URL https://www.sciencedirect.com/science/article/pii/S1319157822001586.

Miao, Q., Xie, L., Cui, H., Liang, W. and Pecht, M. 2013. Remaining useful life prediction of lithium-ion battery with unscented particle filter technique. Microelectronics Reliability, 53(6): 805–810. ISSN 0026-2714. URL https://www.sciencedirect.com/science/article/pii/S0026271412005239.

Patil, B. P. 2023. Accelerating AI-based battery management system's SoC and SoH on FPGA. 2023, Jun 2023.

Patil, M. A., Tagade, P., Hariharan, K. S., Kolake, S. M., Song, T., Yeo, T. and Doo, S. 2015. A novel multistage support vector machine based approach for li-ion battery remaining useful life estimation. Applied Energy, 159: 285–297. ISSN 0306-2619. URL https://www.sciencedirect.com/science/article/pii/S0306261915010557.

Probst, P., Wright, M. N. and Boulesteix. A. -L. 2019. Hyperparameters and tuning strategies for random forest. WIREs Data Mining and Knowledge Discovery, 9(3): e1301. URL https://wires.onlinelibrary.wiley.com/doi/abs/10.1002/widm.1301.

Pushkar, K., Napa, N., Madichetty, S., Agrawal, M. K. and Tamma, B. M. 2022. Thermal analysis of lithium ion battery pack with different cooling media. Technical Report, SAE Technical Paper.

Qaisar, S. M. 2020. Event-driven approach for an efficient coulomb counting based li-ion battery state of charge estimation. Procedia Computer Science, 168: 202–209. ISSN 1877-0509. URL https://www.sciencedirect.com/science/article/pii/S1877050920304075.

Qaisar, S. M. and AbdelGawad. A. E. E. 2021. Prediction of the li-ion battery capacity by using event-driven acquisition and machine learning. In 2021 7th International Conference on Event-Based Control, Communication, and Signal Processing (EBCCSP), pp. 1–6.

Qaisar, S. M. and Aljefri, R. 2021. Adaptive-rate method for the power quality disturbances identification in the 5g framework. Procedia Computer Science, 182: 115–120. ISSN 1877-0509. URL https://www.sciencedirect.com/science/article/pii/S1877050921004804.

Qaisar, S. M. and AlQathami, M. 2021. Event-driven sampling based li-ion batteries soh estimation in the 5g framework. Procedia Computer Science, 182: 109–114, ISSN 1877-0509. URL https://www.sciencedirect.com/science/article/pii/S1877050921004798.

Qaisar, S. M. and Alsharif, F. 2020. Event-driven system for proficient load recognition by interpreting the smart meter data. Procedia Computer Science, 168: 210–216. ISSN 1877-0509. URL https://www.sciencedirect.com/science/article/pii/S1877050920304063.

Qaisar, S. M. and Alyamani, N. 2022. A Review of Charging Schemes and Machine Learning Techniques for Intelligent Management of Electric Vehicles in Smart Grid, pp. 51–71. Springer International Publishing, Cham. ISBN 978-3-030-93585-6.

Qaisar, S. M., Alzahrani, W., Almojalid, F. M. and Hammad, N. 2019a. A vehicle movement based self-organized solar powered street lighting. In 2019 IEEE 4th International Conference on Signal and Image Processing (ICSIP), pp. 445–448.

Qaisar, S. M., Laskar, S., Lunglmayr, M., Moser, B. A., Abdulbaqi, R. and Banafia, R. 2019b. An event-driven approach for time-domain recognition of spoken english letters. In 2019 5th International Conference on Event-Based Control, Communication, and Signal Processing (EBCCSP), pp. 1–4.

Qaisar, S. M., Khan, S. I., Dallet, D., Tadeusiewicz, R. and lawiak. P. P. 2022. Signal-piloted processing metaheuristic optimization and wavelet decomposition based elucidation of arrhythmia for mobile healthcare. Biocybernetics and Biomedical Engineering, 42(2): 681–694. ISSN 0208-5216. URL https://www.sciencedirect.com/science/article/pii/S0208521622000444.

Quynh, N. V., Ali, Z. M., Alhaider, M. M., Rezvani, A. and Suzuki, K. 2021. Optimal energy management strategy for a renewable-based microgrid considering sizing of battery energy storage with control policies. International Journal of Energy Research, 45(4): 5766–5780. URL https://onlinelibrary.wiley.com/doi/abs/10.1002/er.6198.

Rana, M., Ishtiaq Hossain Khan, M., Nshizirungu, T., Jo, Y. -T. and Park, J. -H. 2023. Green and sustainable route for the efficient leaching and recovery of valuable metals from spent ni-cd batteries. Chemical Engineering Journal, 455: 140626. ISSN 1385-8947. URL https://www.sciencedirect.com/science/article/pii/S138589472206106X.

Raoofi, T. and Yildiz, M. 2023. Comprehensive review of battery state estimation strategies using machine learning for battery management systems of aircraft propulsion batteries. Journal of Energy Storage, 59: 106486. ISSN 2352-152X. URL https://www.sciencedirect.com/science/article/pii/S2352152X22024756.

Ren, L., Zhao, L., Hong, S., Zhao, S., Wang, H. and Zhang, L. 2018. Remaining useful life prediction for lithium-ion battery: A deep learning approach. IEEE Access, 6: 50587–50598.

Ren, Z. and Du, C. 2023. A review of machine learning state-of-charge and state-of-health estimation algorithms for lithium-ion batteries. Energy Reports, 9: 2993–3021. ISSN 2352-4847. URL https://www.sciencedirect.com/science/article/pii/S235248472300118X.

Roberts, B. P. and Sandberg, C. 2011. The role of energy storage in development of smart grids. Proceedings of the IEEE, 99(6): 1139–1144.

Sabo, A., Qaisar, S. M., Subasi, A. and Rambo, K. A. 2018. An event driven wireless sensors network for monitoring of plants health and larva activities. In 2018 21st Saudi Computer Society National Computer Conference (NCC), pp. 1–7.

Saha, B., Poll, S., Goebel, K. and Christophersen, J. 2007. An integrated approach to battery health monitoring using bayesian regression and state estimation. In 2007 IEEE Autotestcon, pp. 646–653.

Sarmah, S., Lakhanlal, S., Kakati, B. K. and Deka, D. 2023. Recent advancement in rechargeable battery technologies. WIREs Energy and Environment, 12(2): e461. URL https://wires.onlinelibrary.wiley.com/doi/abs/10.1002/wene.461.

Schneider, D. and Endisch, C. 2020. Robustness and reliability of model-based sensor data fusion in a lithium-ion battery system. In 2020 IEEE Conference on Control Technology and Applications (CCTA), pp. 685–691.

Shen, D., Wu, L., Kang, G., Guan, Y. and Peng, Z. 2021. A novel online method for predicting the remaining useful life of lithium-ion batteries considering random variable discharge current. Energy, 218: 119490. ISSN 0360-5442. URL https://www.sciencedirect.com/science/article/pii/S0360544220325974.

Shu, X., Shen, S., Shen, J., Zhang, Y., Li, G., Chen, Z. and Liu, Y. 2021. State of health prediction of lithium-ion batteries based on machine learning: Advances and perspectives. iScience, 24(11): 103265. ISSN 2589-0042. URL https://www.sciencedirect.com/science/article/pii/S2589004221012347.

Singh, S. S. S., Rezaei, S. and Birke, K. P. 2023. Hybrid modeling of lithium-ion battery: Physics-informed neural network for battery state estimation. 9(6), May 2023.

Song, Y., Li, L., Peng, Y. and Liu, D. 2018. Lithium-ion battery remaining useful life prediction based on gru-rnn. In 2018 12th International Conference on Reliability, Maintainability, and Safety (ICRMS), pp. 317–322.

Sun, C., Lin, H., Cai, H., Gao, M., Zhu, C. and He, Z. 2021. Improved parameter identification and state-of-charge estimation for lithium-ion battery with fixed memory recursive least squares and sigma-point kalman filter. Electrochimica Acta, 387: 138501. ISSN 0013-4686. URL https://www.sciencedirect.com/science/article/pii/S001346862100791X.

Sun, F., Xiong, R. and He, H. 2016. A systematic state-of-charge estimation framework for multi-cell battery pack in electric vehicles using bias correction technique. Applied Energy, 162: 1399–1409. ISSN 0306-2619. URL https://www.sciencedirect.com/science/article/pii/S0306261914012707.

Tang, S., Yu, C., Wang, X., Guo, X. and Si, X. 2014. Remaining useful life prediction of lithium-ion batteries based on the wiener process with measurement error. Energies, 7(2): 520–547. ISSN 1996-1073. URL https://www.mdpi.com/1996-1073/7/2/520.

Tao, L., Zhang, T., Hao, J., Wang, X., Lu, C., Suo, M. and Ding, Y. 2020. Degradation dynamics cognition and prediction of li-ion battery: An integrated methodology for alleviating range anxiety. IEEE Acccss, 8: 183927–183938.

Tong, S., Lacap, J. H. and Park, J. W. 2016. Battery state of charge estimation using a load-classifying neural network. Journal of Energy Storage, 7: 236–243. ISSN 2352-152X. URL https://www.sciencedirect.com/science/article/pii/S2352152X16300949.

Wang, D., Miao, Q. and Pecht, M. 2013. Prognostics of lithium-ion batteries based on relevance vectors and a conditional three-parameter capacity degradation model. Journal of Power Sources, 239: 253–264. ISSN 0378-7753. URL https://www.sciencedirect.com/science/article/pii/S0378775313005235.

Wang, Z., Feng, G., Zhen, D., Gu, F. and Ball, A. 2021. A review on online state of charge and state of health estimation for lithium-ion batteries in electric vehicles. Energy Reports, 7: 5141–5161. ISSN 2352-4847. URL https://www.sciencedirect.com/science/article/pii/S2352484721007150.

Xiong, R., Cao, J., Yu, Q., He, H. and Sun, F. 2018. Critical review on the battery state of charge estimation methods for electric vehicles. IEEE Access, 6: 1832–1843.

Xuan, L., Qian, L., Chen, J., Bai, X. and Wu, B. 2020. State-of-charge prediction of battery management system based on principal component analysis and improved support vector machine for regression. IEEE Access, 8: 164693–164704.

Yang, G., Ma, Q., Sun, H. and Zhang, X. 2022. State of health estimation based on gan-lstm-tl for lithium-ion batteries. International Journal of Electrochemical Science, 17(11): 221128. ISSN 1452-3981. URL https://www.sciencedirect.com/science/article/pii/S145239812302655X.

Yang, K., Zhang, L., Zhang, Z., Yu, H., Wang, W., Ouyang, M., Zhang, C., Sun, Q., Yan, X., Yang, S. and Liu, X. 2023. Battery state of health estimate strategies: From data analysis to end-cloud collaborative framework. Batteries, 9(7). ISSN 2313-0105. URL https://www.mdpi.com/2313-0105/9/7/351.

You, M., Liu, Y., Chen, Z. and Zhou, X. 2022. Capacity estimation of lithium battery based on charging data and long short-term memory recurrent neural network. In 2022 IEEE Intelligent Vehicles Symposium (IV), pp. 230–234.

Zhang, D., Zhong, C., Xu, P. and Tian, Y. 2022. Deep learning in the state of charge estimation for li-ion batteries of electric vehicles: A review. Machines, 10(10). ISSN 2075-1702. URL https://www.mdpi.com/2075-1702/10/10/912.

Zhang, H., Miao, Q., Zhang, X. and Liu, Z. 2018a. An improved unscented particle filter approach for lithium-ion battery remaining useful life prediction. Microelectronics Reliability, 81: 288–298. ISSN 0026-2714. URL https://www.sciencedirect.com/science/article/pii/S0026271417306005.

Zhang, S., Zhai, B., Guo, X., Wang, K., Peng, N. and Zhang, X. 2019. Synchronous estimation of state of health and remaining useful lifetime for lithium-ion battery using the incremental capacity and artificial neural networks. Journal of Energy Storage, 26: 100951. ISSN 2352-152X. URL https://www.sciencedirect.com/science/article/pii/S2352152X19307340.

Zhang, W. and Wang, W. 2018. Lithium-ion battery soc estimation based on online support vector regression. In 2018 33rd Youth Academic Annual Conference of Chinese Association of Automation (YAC), pp. 564–568.

Zhang, Y., Xiong, R., He, H. and Pecht, M. G. 2018b. Long short-term memory recurrent neural network for remaining useful life prediction of lithium-ion batteries. IEEE Transactions on Vehicular Technology, 67(7): 5695–5705.

Zhao, D., Li, H., Zhou, F., Zhong, Y., Zhang, G., Liu, Z. and Hou, J. 2023. Research progress on data-driven methods for battery states estimation of electric buses. World Electric Vehicle Journal, 14(6). ISSN 2032-6653. URL https://www.mdpi.com/2032-6653/14/6/145.

Zheng, X. and Fang, H. 2015. An integrated unscented kalman filter and relevance vector regression approach for lithium-ion battery remaining useful life and short-term capacity prediction. Reliability Engineering & System Safety, 144: 74–82. ISSN 0951-8320. URL https://www.sciencedirect.com/science/article/pii/S0951832015002148.

CHAPTER 7

Addressing Societal Challenges and Enhancing Academic Effectiveness through Challenge-Driven Education: A Case Study in Smart Microgrid

Diana Rwegasira,[a,*] *Imed Ben Dhaou,*[b,c,d] *Hellen Maziku*[a]
and *Hannu Tenhunen*[e]

1. Introduction

In today's society, engineering education plays an indispensable role in driving innovation, technological advancements, and economic prosperity. Engineers serve as the driving force behind technological innovation, transforming groundbreaking ideas into tangible solutions that shape our world. To keep pace with the ever-evolving technological landscape and equip engineers with the skills necessary to address the challenges and opportunities of the 21st century, the engineering curriculum has undergone significant transformations. As highlighted in numerous educational reports, industry, particularly in the context of Industry 4.0, demands engineers who possess not only technical expertise but also a repertoire of transferable skills, including communication, teamwork, lifelong learn-

[a] Department of Computer Science and Engineering, College of ICT, University of Dar es salaam, P.O.BOX 33335, Tanzania.

[b] Department of Computer Science, Hekma School of Engineering, Computing, and Design, Dar Al-Hekma University, Jeddah 22246-4872, Saudi Arabia.

[c] Department of Computing, University of Turku, Finland.

[d] Department of Technology, Higher Institute of Computer Sciences and Mathematics, University of Monastir, Tunisia.

[e] Faculty of Information Technology and Communication Tampere University, UC Pori, Finland.

Email: imed.bendhaou@utu.fi

* Corresponding author: dianarosetz@gmail.com

ing, leadership, creativity, critical thinking, and entrepreneurial acumen (Ekren and Kumar, 2020; Lingard and Barkataki, 2011). Expected learning outcomes (ELOs) serve as formal declarations of a learner's ability to demonstrate knowledge, skills, and attitudes at the culmination of a learning process.

The development of an innovative engineering program begins with the definition of the Program Learning Outcomes (PLOs) or student outcomes, as outlined by ABET (Accreditation Board for Engineering and Technology). These PLOs should be in line with the Institutional Learning Outcomes (ILOs) and the National Qualification Framework (NQF). The PLOs are then used to create the curriculum, which consists of a large number of courses designed to provide students with the knowledge, skills, and values required by the PLOs. Each course should have its own Course Learning Outcomes (CLOs) that must be assigned to the PLOs (Felder and Brent, 2003).

Faculty development assumes a pivotal role in shaping the design and execution of academic programs. Within any department, school, or institution, a dedicated professional development unit or program should be committed to enhancing faculty members' expertise in crucial areas, including pedagogy, teaching methods, technology integration, and assessment strategies. In the context of engineering programs, it becomes imperative for faculty members not only to excel in these foundational areas but also to actively cultivate collaborative relationships with industry professionals. This synergy ensures a dynamic and practical educational experience for students, aligning academic pursuits with real-world industry needs. To address these industry needs effectively, the inclusion of specialized and capstone-oriented courses is recommended, as suggested in the work by (Dhaou, 2022).

According to (Dixson and Worrell, 2016), both formative and summative assessment approaches are essential for effectively measuring learning outcomes and fostering continuous improvement in the learning process. Recognizing the widening gap between engineering education and industry expectations, MIT developed the CDIO (Conceive, Design, Implement, and Operate) framework to bridge this gap and prepare engineers for success in the real world (Crawley, 2002). The CDIO curriculum encompasses four core components: (a) technical knowledge and reasoning, (b) human and professional abilities and traits, (c) interpersonal skills, and (d) the ability to conceive, create, implement, and operate systems within a company and for social purposes. Over 140 universities and higher education institutions worldwide have adopted the CDIO framework, demonstrating its widespread recognition and effectiveness in preparing engineers for the demands of the 21st century.

CDE methodology, on the other hand, is a new frontier in engineering education that aims to incorporate societal requirements into university teaching through projects while also increasing engineering creativity. In this paradigm, a group of students with interdisciplinary skills works on societal and/or industry-related problems. CDE allows Engineering Education programs to include Cognitive learning (knowledge), Psychomotor learning (skills), and affective learning (attitude). Students improve their logical, intuitive, and creative thinking as well as interpersonal skills such as awareness, communication, and cooperation as they work in groups and interact with challenge stakeholders. As CDE aspires to provide society with long-term and practical solutions, it exposes students to practical skills such as manual dexterity and the use of methodologies, materials, tools, and instruments.

The Royal Institute of Technology (KTH-Sweden) has advocated for the CDE methodology in order to design graduate courses that solve societal problems through engaging, multidisciplinary, and open-ended teams. To meet the challenges of the single market economy and maintain a sustainable growth rate for modern society, universities should transition from a learning center to an innovation center. Initially, the triple helix model was promoted as a means of fostering innovation (Leydesdorff, 2000). Society and media were recently added as the fourth quadrant (McAdam and Debackere, 2018). Figure 1 depicts the quadruple helix model as well as the CDE quadrants.

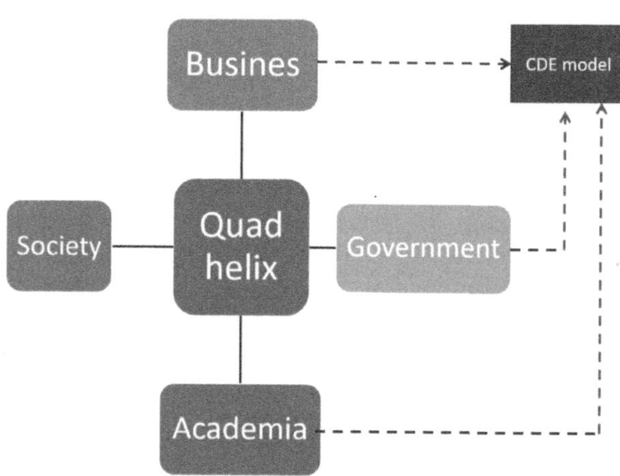

Figure 1. Quadruple Helix model of innovation.

2. Background

2.1. The challenge driven education framework

The concept of challenge-driven education (CDE) can be traced back to the early 2000s, when researchers at the Swiss Federal Institute of Technology Lausanne (EPFL) began to explore ways to make learning more relevant and engaging for students. The researchers were inspired by the success of project-based learning (PBL) and other hands-on learning approaches, but they wanted to develop a model that would go beyond simply giving students projects to work on.

The EPFL researchers envisioned a model of education where students would be engaged in real-world challenges that would require them to use their knowledge and skills to solve complex problems. This model would be based on the following principles:

Authenticity: Students would work on real-world challenges that are relevant to their interests and to the world around them. Interdisciplinarity: Students would draw on knowledge and skills from multiple disciplines to solve the challenges they face. Collaboration: Students would work together in teams to solve the challenges, developing their teamwork and collaboration skills in the process. Reflection: Students would reflect on their learning throughout the process, identifying their strengths and weaknesses and developing new strategies for solving problems. The EPFL researchers began to implement CDE in their own courses and found that it was a very effective way to engage students and help them to learn deeply. CDE has since been adopted by educators around the world, and it is now used in a variety of educational settings, from primary schools to universities.

Here are some key milestones in the development of CDE: In 2003, EPFL researchers initiated the idea of Challenge-Driven Education (CDE). Five years later, the first CDE course was offered at EPFL. In 2010, the Challenge-Based Learning Research Center was established at the same institution. In 2012, the first international conference on challenge-based learning was held. Three years later, the Challenge-Based Learning Network was launched. By 2020, CDE had become a popular teaching method in schools and universities worldwide.

CDE is a relatively new approach to education, but it is quickly gaining popularity as educators recognize the benefits that it can offer to students. CDE is a way to make learning more relevant, engaging, and effective. It is a way to prepare students for the challenges of the 21st century. CDE is a methodology that incorporates active learning throughout the process, from identifying the problem to providing a solution. At every stage of its application, the approach connects research, education, and innovation. The approach involves students, community/stakeholders, and researchers working together to provide a service or solution to a specific problem. Research done by (Magnell and Högfeldt, 2015) describes the CDE approach and how to implement it in practice. CDE has several advantages, including the ability to solve societal problems, and to work in a multidisciplinary team work to bridge the gap between academic and societal needs. The process involved in the CDE approach is depicted in Figure 2.

Figure 2. From identification of the problem to solution deployment (Ebrahimi et al., 2019).

2.2. Adopting challenge-driven education in engineering program

In 2015, the University of Dar es Salaam (UDSM) and the Royal Institute of Technology (KTH) joined forces to launch the iGrid project, which was funded by SIDA. The primary objective of the project was to create and upgrade smart grid capacity in Tanzania. This had an impact on both the main grid and the microgrid. TANESCO, the main shareholder of the main grid, set three challenges for MSc and PhD students to tackle: (i) Inefficient power system fault prevention and clearance (C1) for PhD students enrolled in 2016; (ii) Lack of power consumption visibility in the electrical secondary distribution network (C2) for MSc students enrolled in 2016; and (iii) Non-optimal management of electrical distribution network due to heterogeneous systems (C3) for MSc students enrolled in 2017.

Furthermore, the iGrid initiative traveled to remote off-grid villages to address the unique electricity difficulties that those populations face. A community in the Kisiju region, 100 kilometers from Dar es Salaam, was visited by the crew. A centrally placed solar-powered microgrid provides electricity to the town, which consists of about 100 homes (Matungwa, 2014). Despite the lack of electrical generation, stakeholders noted two major challenges:

Manual and inefficient control of power distribution (C4), for Ph.D. student enrolled in the year 2016. Lack of stable and secure communication infrastructure that can support autonomous and remote control and monitoring (C5), for PhD students enrolled in the year 2016. Students from several disciplines were enrolled in trans-disciplinary teams. Information systems, telecommunications, computer engineering, data science, electronics, and computer science are all examples of specializations. Table 1 outlines the primary challenges and sub-challenges undertaken by students, as well as the categorization of the problem to be solved. The assigned number of graduate students for each challenge is summarized in Table 2. Figures 3 and 4 detail the project work undertaken by PhD and MSc students, respectively, as well as their disciplines and implementation phases. The achievement level was benchmarked using the five implementation phases of (Rådberg et al., 2020). These phases are: (1) problem formulation; (2) idea or model generation; (3) concept development; (4) testing/evaluation within an academic setting; (5) testing/evaluation by external stakeholders.

2.3. Contributions of this study

This research study focuses on assessing the use of the CDE approach at the University of Dar es Salaam (UDSM) and the Royal Institute of Technology (KTH) to demonstrate its impact on society. The problem solved was based on how to provide a reliable and sustainable solution for power generation in Tanzania, for both rural and urban areas. The case study can be used to cover different topics that have been overlooked in CDE, in particular the problem-solving process, lifelong learning, and team dynamics.

Table 1.   Task distribution using CDE approach.

Description	Sub-challenges
C1: Inefficient power system fault prevention and clearance	Detection and restoration Monitoring, control and prevention Management systems and services Support infrastructures and frameworks
C2: Lack of power consumption visibility	Data acquisition Data transmission Data storage Data analytics Data visualization
C3:Non-optimal management of electrical distribution network due to heterogeneous systems	Integration of GIS and AMI Fault analysis Load forecasting Outage management Theft detection
C4:Manual and inefficient control of power distribution in rural areas (microgrid)	Agent based system for control and monitoring
C5:Lack of stable and secure communication in rural areas (microgrid)	Secure system that can support autonomous and remote control and monitoring

Table 2.   Distribution of graduate students among the five challenges.

Challenge	Students
C1	Eight PhD students enrolled at UDSM in the year 2016
C2	Six master students enrolled at UDSM in the year 2016
C3	Five master students enrolled at UDSM in the year 2017
C4	One student enrolled in a co-tutelle PhD program (UDSM and KTH)
C5	One student enrolled in a co-tutelle PhD program (UDSM and KTH)

3. Methods

This study employs a mixed-methods approach, combining quantitative and qualitative research methods, to evaluate the effectiveness of the CDE approach. A total of 50 participants, including 30 students, 10 instructors, and 10 stakeholders, were invited to complete a comprehensive questionnaire. Of the 50 participants, 39 responded, resulting in an approximate 78% response rate. The respondents included 24 students, 8 instructors, and 7 stakeholders. The questionnaire was designed to assess the skills acquired through the CDE approach, evaluate the overall learning experience, and identify areas for improvement. The survey was conducted after the majority of participants had completed their studies to gather more comprehensive feedback on the CDE approach. Participants were asked to rate each statement on a five-point Likert scale ranging from "Strongly Agree" to "Strongly Disagree" or "Very Easy" to "Very Difficult." Additionally, open-ended questions were included to allow participants to provide detailed explanations and elaborations on their responses. The questionnaires are attached in the Appendix A for reference (Tables A1 4 5).

Year	Title of Project	Discipline	CDE Implementation Phase				
			1	2	3	4	5
	Hybrid communication architecture for automatic fault detection and clearance in Electrical Power Secondary Distribution Network based on distributed processing	Telecom	•	•	•	•	•
	Optimal Sensor Network and Faults Classification Algorithms for Automatic Faults Clearance in Electrical Secondary Distribution Network	CE	•	•	•	•	•
	Integrated Architectural Model for enhancing fault clearance applications in Electrical Secondary Distribution Network	CS	•	•	•	•	•
	Distributed Algorithm for Enhanced Fault Localization and Service Restoration to Improve Reliability in Electrical Secondary Distribution Network	Telecom	•	•	•	•	•
	Distributed Energy Resources Placements and Coordination Algorithms for Enhancing Service Restoration in The Electrical Secondary Distribution Network	Electronics	•	•	•	•	•
2016	Cross Layers Resilient Communication Network for Fault Detection and Clearance Automation in Secondary Distribution Power Grid	CE	•	•	•	•	•
	Agent Based System for Enhanced Controlling and Monitoring of a Solar Driven DC Microgrid	CE	•	•	•	•	•
	Developing a Security-Enhanced Internet-of-Things Based Communication System for Smart Microgrid	CE	•	•	•	•	•
	A Real-Time Condition-Based Equipment Failure Prediction for Maintenance Decision Support in Secondary Distribution Network	Electronics	•	•	•	•	
	Efficient Scheme for Secured Communication Network for The Internet of Things Enabled Distribution Automation	CS	•	•	•	•	
	A Parallel Big Data-Based Algorithm for Fault Prediction in Electrical Secondary Distribution Network.	CS	•	•	•	•	
2017	Event-Driven Load Demand Prediction Model with Outlier Removal Using Unsupervised Machine Learning for Power Management in Automatic Fault Clearance at Secondary Electric Power Distribution Network	CS	•	•	•	•	
2019	Algorithm Development for Optimal Adjustment of Relays Parameters by Time and Current Characteristics in a Secondary Distribution Network and Relays	EE	•	•	•	•	

Figure 3. Evaluating the contributions of CDE projects undertaken by PhD students.

4. Results

The CDE has fostered teamwork abilities, improved communication skills, and boosted student confidence. However, the amount of work completed in this course seems to be disproportionate to the number of credits earned.

Tables 3, 4, 5 present the survey results, which encompass perspectives from students, instructors, and stakeholders.

5. Discussions

The analysis of the impact of CDE practices uses an Information Systems theoretical approach which focusing on the following parameters: (i) Expectations (effort and performance), (ii) Usefulness, (iii) Satisfaction, (iv) Ease of Use, and (v) Facilitating Conditions. These are the Unified Theory of Acceptance and Use of Technology (UTAUT) model's Information System (IS) by (Venkatesh et al., 2016).

Year	Title of Project	Students' Discipline	CDE Implementation Phase				
			1	2	3	4	5
2016	Development of a Scalable Big Data Visualizer for Power Consumption in the Electrical Secondary Distribution Secondary Network	Computer Science	•	•	•	•	•
	Adaptation of Internet of Things Architecture on Smart Meter to facilitate Interoperability Across Advanced Metering Infrastructure	Computer Engineering	•	•	•	•	•
	Low-Cost Acquisition of Power Consumption Data in Electrical Secondary Distribution Secondary Network	Computer Engineering	•	•	•	•	•
	Development of Big Data Analytics System to Enhance Visibility of Power Consumption in the Electrical Secondary Distribution Secondary Network	Computer Science	•	•	•	•	
	Big Data Modeling for Managing Heterogeneous Data from Smart Grid Secondary Electrical Distribution Network in Tanzania	Computer Science	•	•	•	•	•
	Development of a Communication Network and Gateway for consumption monitoring in Secondary Distribution of Electrical power grid.	Computer Engineering	•	•			
2017	Developing an Algorithm for Enhancing Long Term Prediction of Distribution Power Transformer Load Using Artificial Neural Networks	Electrical Engineering	•	•	•	•	
	Development of Model-Based Data Driven Real Time Fault Analysis Tool to Improve Fault Detection Accuracy in Electrical Secondary Distribution Secondary Network	Computer Engineering	•	•	•	•	•
	Develop an Algorithm to Estimate Outage Location of An Electrical Distribution Network	Computer Engineering	•	•	•	•	
	Development of Data Analytic Tool to Enhance Power Theft Detection in Electrical Power Distribution Network	Computer Engineering	•	•	•		
	Designing and Developing a Middleware layer Service to integrate AMI and GIS to Facilitate the Management of Low Voltage Network by Feeding GIS with Real-time Data from Field Services	Computer Engineering	•	•	•	•	•

Figure 4. Evaluating the contributions of CDE projects undertaken by master students.

Table 3. Results from the students' survey.

Question	Strongly Disagree	Disagree	Neutral	Agree	Strongly Agree
Q1	0%	0%	0%	57.1%	42.9%
Q2	0%	0%	35.7%	35.7%	28.6%
Q3	0%	0%	0%	35.7%	57.1%
Q4	0%	0%	0%	35.7%	64.3%
Q5	0%	14.3%	28.6%	57.1%	0%
Q6	0%	0%	0%	57.1%	42.9%
Q7	0%	7.1%	21.4%	42.9%	28.6%
Q8	0%	0%	14.3%	64.3%	21.4%
Q9	7.1%	0%	14.3%	50%	35.7%
Q10	0%	0%	14.3%	42.9%	42.9%
Q11	0%	0%	14.3%	50%	35.7%

5.1. Expectation

Performance Expectancy (PE), a key component of the UTAUT instrument, is the degree to which people believe a technology or innovation will improve their performance. The perceived ease of use is referred to as Effort Expectancy (EE). The survey questions were designed to determine whether CDE, as a learn-

Table 4.　Faculty evaluation.

Question	Strongly Disagree	Disagree	Neutral	Agree	Strongly Agree
Q1	16.7%	0%	16.7%	33.3%	33.3%
Q2	0%	0%	0%	33.3%	66.7%
Q3	0%	0%	16.7%	50%	33.3%
Q4	0%	0%	33.3%	33.3%	33.3%
Q5	16.7%	0%	16.7%	33.3%	33.3%
Q6	0%	0%	33.3%	33.3%	33.3%
Q7	0%	0%	0%	50%	50%
Q8	0%	0%	0%	50%	50%

Table 5.　Stakeholder's feedback.

Question	Strongly Disagree	Disagree	Neutral	Agree	Strongly Agree
Q1	0%	33.3%	0%	66.7%	0%
Q2	0%	0%	0%	66.7%	33.3%
Q3	0%	0%	0%	33.3%	66.7%
Q4	0%	0%	0%	100%	0%
Q5	0%	0%	0%	33.3%	66.7%
Q6	0%	0%	66.7%	33.3%	0%
Q7	0%	0%	33.3%	33.3%	33.3%
Q8	0%	33.3%	0%	66.7%	0%

ing and problem-solving methodology, performed at the level expected by the participants. For example, more than half of the student respondents (57.1%) and instructors (66.6%) agreed that the effort put in delivering the course met their expectations and was aligned with the designed curriculum. The process of problem solving for the challenge supplied followed the expected methodology phases of problem definition, solution development, course of action decision, solution implementation, and evaluation, according to a large proportion of students (57.1%) or strongly agreed (42.9%). 83.3% of teachers believed that knowledge delivered to students was evaluated through the use of both summative and formative assessment grading criteria. It's worth noting that all of the stakeholder respondents (100%) want people/investors to pay for the solutions that have been implemented.

The survey results revealed that 66.7% of stakeholders had attended other CDE-based workshops with other institutions, indicating that they were aware of CDE as a problem-solving methodology. This demonstrates CDE's acceptability and potential adoption as a primary tool for learning and problem solving in a variety of institutions.

5.2. Usefulness

The survey results highlighted the impact of the CDE on students, instructors and stakeholders. For about 57.1%, students expressed the usefulness of the CDE on improving their communication skills and teamwork (soft skills). CDE provided students further understanding of the 17 UN sustainable goals where by at the beginning majority did not have that knowledge. This can be seen as only 35% agree to have the idea about it. Also, 64% of students were able to get new skills in entrepreneur through the use of CDE approach. CDE was also useful in consolidating and improving the leadership, teamwork and communication skills of the instructors. Last not least, 66% of the stakeholders expressed the usefulness of the CDE in addressing the societal needs and delivering sustainable solution.

5.3. Satisfaction

The level of satisfaction is typically evaluated based on a variety of factors, such as student grades, instructor expectations, and the product desired by stakeholders. This study found that 84 percent of students were content with the course design, delivery method, and grade they received. Additionally, while half of the instructors were satisfied with their students' presentation, the other half were not due to its time-consuming nature. Moreover, the perspectives of stakeholders were considered in terms of comparing this approach to prior ones. When using the CDE approach, 67 percent of stakeholders agreed that the risk is minimal since meetings must be held at each stage, allowing for close supervision and reducing the chances of failure.

5.4. Ease of use

Many studies have shown that students often struggle to understand and solve problems. The CDE approach, however, has been found to be beneficial; 57.1% of students reported that it was easy to understand the project and felt comfortable when designing and executing solutions. The desired learning outcomes were also clearly visible in the output, which clarified the concept learning theories. Additionally, 85% of stakeholders strongly agreed with the approach, demonstrating its effectiveness. Furthermore, students found that the solutions were straightforward and easy to comprehend.

5.5. Facilitation conditions

Consideration of facilitation conditions is crucial when evaluating the CDE approach. This approach ensures that students are guided in a manner that is equitable and conducive to grasping the discipline-specific knowledge, skills, and mindsets essential for course completion. Impressively, 67% of instructors strongly affirmed the effectiveness of this approach, underscoring its strengths in the SWOT analysis of the CDE. Moreover, involving stakeholders throughout the course streamlined engagement and commitment. A substantial 66.7% of respondents expressed strong agreement that facilitation conditions, such as providing constructive feedback and fostering discussions with students, are pivotal and practical aspects of the CDE approach.

6. Harnessing the potential of CDE approach

6.1. Important aspects to consider upon using CDE

Three key points have been raised by respondents in terms of improving the CDE approach. These are:

(1) Students' level of education: It is important to consider the students' experience and level of education as this will also help with the project's completion. Students with more experience and education will be better equipped to face the challenges of the CDE approach.

(2) Site visits: Site visits were mentioned as the first input to consider/implement when dealing with CDE. This will assist students in having a clear picture of what they are supposed to solve rather than guessing. Site visits will also help students to develop a better understanding of the customer's needs.

(3) Funds for designing the prototype: Once delivered to the customer, it is also important to have a complete prototype to demonstrate it. This also adds value and credibility to the students. As a result, having funds for implementation increases the potential of the prototypes/products created.

Numerous respondents did not prioritize the curriculum, placing greater emphasis on meetings and stakeholder feedback. While acknowledging the importance of enhancing the curriculum, there is a consensus that these improvements should be implemented promptly without undue delay.

In addition to the three key points raised by respondents, the following are some additional important aspects to consider upon using CDE:

- Clearly defined project goals and objectives: The project goals and objectives should be clearly defined and communicated to all stakeholders. This will help to ensure that everyone is on the same page and that the project is on track to meet its objectives.
- Effective communication and collaboration: Effective communication and collaboration are essential for the success of any CDE project. Students, instructors, and stakeholders must be able to communicate effectively with each other in order to share ideas, solve problems, and make decisions.
- Regular feedback and evaluation: Regular feedback and evaluation are important for ensuring that the CDE project is on track and that students are learning effectively. Students should receive feedback on their work from instructors and stakeholders on a regular basis.
- Flexibility and adaptability: CDE projects are often complex and dynamic, and it is important to be flexible and adaptable in order to deal with changes and challenges. Students, instructors, and stakeholders must be willing to adapt their plans and expectations as needed.

6.2. *Monetizing solution process*

The iGrid projects presented in this paper concentrate on solutions developed to address sub-challenges resulting from the larger challenge breakdown utilizing the CDE approach. Furthermore, as shown in Table 1, the majority of these solutions are undergoing testing and review by external stakeholders. It is critical that these ideas be combined to create a single solution that addresses the larger problem holistically. The steps that must be taken to ensure that the solution is not only monetized but also adopted in a long-term manner include:

- a) Integration of sub-challenge solutions.
- b) Technical Testing and evaluation of the main solution.
- c) Engagement of stakeholders and industry in testing and evaluation.
- d) Alignment with government policies and initiatives.
- e) Feasibility analysis of pilot deployment and related costs.
- f) Demonstration of pilot findings to attract solution adopters/users.
- g) Solution use through startups, Labs, purchase, partnerships, etc.
- h) CDE based improvement of existing solutions.

6.3. *Way forward regarding CDE approach*

It is imperative for every institution to prioritize the appreciation of student ideas and to actively involve stakeholders right from the project's inception. A respondent highlighted that the Collaborative Design and Engineering (CDE) approach has demonstrated efficacy in addressing and resolving prevalent socioeconomic issues. Moreover, to yield tangible output, it is recommended to conduct site visits at the project's commencement rather than waiting until the mid-way point of the students' studies.

The stakeholders from TANESCO, who play a pivotal role in the CDE, endorse the value, innovativeness, and sustainability of the students' solutions, emphasizing their relevance to society. Additionally, they propose that students engage in visits to utility companies to gain familiarity with real-world challenges. This proactive approach ensures that students are well-versed in practical considerations from the outset of their projects.

7. Conclusion

In the collaborative nexus of academic institutions, businesses, and governments, challenge-driven education emerges as a promising framework for addressing societal needs. The participation of graduate students in Phase 1 of the Challenge-Driven Education (CDE) has proven instrumental, providing them with invaluable insights into authentic engineering problems while fostering crucial teamwork and communication skills. Despite these achievements, there is a need to strengthen entrepreneurial skills, refine systematic problem-solving processes, and improve the integration of civil society within the CDE framework.

Noteworthy achievements include the timely and collaborative completion of student projects during the period from October 2021 to February 2022. Of these, 8 PhD students successfully concluded their projects, while 6 out of 11 Master's projects navigated through all five phases. The rigorously tested and validated solutions across the majority of projects within academic contexts signify the tangible effectiveness of the CDE approach. This success underscores the importance of active stakeholder participation in the design and implementation of solutions for societal challenges. Moving forward, a greater emphasis on entrepreneurial experience and a more seamless integration of civil society perspectives are pivotal to further enhancing the overall efficacy and impact of the CDE approach.

Acknowledgments

This work has been supported by the Swedish International Development Agency, SIDA.

References

Crawley, E. F. 2002. Creating the cdio syllabus, a universal template for engineering education. volume 2.

Dhaou, I. B. 2022. Aligning engineering education with industrial needs through specialized courses. In 2022 IEEE Global Engineering Education Conference (EDUCON), pp. 1664–1669.

Dixson, D. D. and Worrell, F. C. 2016. Formative and summative assessment in the classroom. Theory into Practice, 55, ISSN 15430421.

Ebrahimi, M., Kelati, A., Nkonoki, E., Kondoro, A., Rwegasira, D., Ben Dhaou, I., Taajamaa, V. and Tenhunen, H. 2019. Creation of cerid: Challenge, education, research, innovation, and deployment "in the context of smart microgrid". In 2019 IST-Africa Week Conference (IST-Africa), pp. 1–8.

Ekren, B. Y. and Kumar, V. 2003. Engineering education towards industry 4.0. volume 0, 2020.

Felder, R. M. and Brent, R. 2003. Designing and teaching courses to satisfy the abet engineering criteria. Journal of Engineering Education, 92(1): 7–25. URL https://onlinelibrary.wiley.com/doi/abs/10.1002/j.2168-9830.2003.tb00734.x.

Leydesdorff, L. 2000. The triple helix: An evolutionary model of innovations. Research Policy, 29. ISSN 00487333.

Lingard, R. and Barkataki, S. 2011. Teaching teamwork in engineering and computer science.

Magnell, M. and Högfeldt, A. -K. 2023. Guide to challenge driven education, 2015. Accessed Nov 21, https://www.kth.se/social/group/guide-to-challenge-d/.

Matungwa, B. J. 2014. An analysis of pv solar electrification on rural livelihood transformation: A case of kisiju-pwani in mkuranga district, tanzania. Master's thesis, University of Oslo. Accessed on 21, 2023. https://www.duo.uio.no/bitstream/handle/10852/43525/1/Masters-Thesis.pdf=.

McAdam, M. and Debackere, K. 2018. Beyond 'triple helix' toward 'quadruple helix' models in regional innovation systems: Implications for theory and practice. R&D Management, 48(1): 3–6. URL https://onlinelibrary.wiley.com/doi/abs/10.1111/radm.12309.

Rådberg, K. K., Lundqvist, U., Malmqvist, J. and Svensson, O. H. 2020. From cdio to challenge-based learning experiences–expanding student learning as well as societal impact? European Journal of Engineering Education, 45. ISSN 14695898.

Venkatesh, V., Thong, J. Y. and Xu, X. 2016. Unified theory of acceptance and use of technology: A synthesis and the road ahead. Journal of the Association for Information Systems, 17. ISSN 15369323.

Appendix A. Questionnaire for graduate students

Following graduation, a questionnaire (see Table A1) was used to evaluate the CDE indirectly. The questionnaire in Table A2 is used to assess the feedback of instructors.

The stakeholders of the iGrid project employed the survey question shown in Table A3 to assess the CDE projects.

Table A1.   Survey questions and corresponding question numbers.

Questions	Q. Number
1. Expectation	
- The effort made in this course is proportional to the number of credits	Q5
- The project covered the following phases: problem definition, solutions generation, course of actions decision, solution implementation, and evaluation	Q6
2. Usefulness	
- This project helped to improve my communication skills	Q1
- I came to know the 17 UN sustainable development goals and helped in solving	Q2
- This project helped me to integrate my skills and reinforced my team working abilities	Q4
- This course is useful to develop entrepreneurial skills	Q7
- The work I did was challenging in a stimulating way	Q8
3. Satisfaction	
- Overall, I am satisfied with the course	Q10
4. Ease of use	
- I understand the challenges, and I feel the project is well-defined. I was able to influence how the challenge is defined and designed in to a solution space	Q3
- What I was expected to learn was clear to me	Q9

Table A2.   Survey questions for faculty members.

Questions	Q. Number
1. Expectation	
- I used both summative and formative assessment grading criteria	Q3
- The amount of student work in this course was proportional to the number of credits gained	Q4
2. Usefulness	
- The project is useful to develop or reinforce leadership skills such as communication, planning, implementation, mindset, teamwork	Q5
3. Satisfaction	
- Overall, I am satisfied with the course	Q2
4. Facilitation Conditions	
- Students have the right: a) Disciplinary knowledge, b) skills, c) mindset to take the course	Q1

Table A3. Survey questions and target participants.

Questions	Q. Number

1. Expectation
    - Do you think that people/investors will be ready to pay for the implemented solution
    - Have you ever attended other workshops with another institution based on CDE approach
	Q6
	
	Q8

2. Usefulness
    - Students' proposed solutions and implementation were valuable to society
    - The solution is innovative and sustainable
	Q2
	
	Q3

3. Ease of Use - The solution is simple and easy to understand	Q4
4. Facilitation Conditions - Meeting engagements and feedback made the implementation easy and realistic	Q5

Part 2

Design and Optimization

CHAPTER 8

Development and Evolution
of Hybrid Microgrids in the Context
# of Contemporary Applications

Yasir Basheer,[a] *Saeed Mian Qaisar*[b,c,]* and *Asad Waqar*[d]

1. Introduction

Energy demand is expanding quickly because of global industrialization and population growth. It is predicted that the amount of energy consumed worldwide would rise by about 50% between 2018 and 2050. The greatest source of energy for meeting the high demand has always been petroleum goods, which has a negative impact on the climate. When petroleum products are burned, a great deal of poisons are released into the air, damaging human health as well as the environment due to the influence of ozone-depleting compounds (Turkdogan, 2021). By lowering atmospheric carbon dioxide (CO_2), which can be accomplished by transitioning to cleaner energy sources, the issue of an increase in Earth's surface temperature may be resolved (Chisale and Mangani, 2021). Envi-

[a] Department of Electrical Engineering, Riphah International University, Islamabad 46000, Pakistan.
[b] CESI LINEACT, Lyon, 69100, France.
[c] Electrical and Computer Engineering Department, Effat University, Jeddah, 22332, Saudi Arabia.
[d] Department of Electrical Engineering, Bahria School of Engineering and Applied Sciences (BSEAS), Bahria University, Islamabad 44000, Pakistan.
* Corresponding author: sqaisar@effatuniversity.edu.sa

ronmentally friendly power sources (RESs), such as solar photovoltaic (PV), hydropower, geothermal, wind, and biomass, might provide everyone, regardless of their geological area, with serious cost options as well as clean and manageable energy (Rehman, 2021). Combining RESs with traditional petroleum derivative-based generators results in half-breed energy frameworks (HESS), which can get over the problem of discontinuity and inconsistent RES supply. Compared to single energy sources, HESS can offer more dependable, manageable, and affordable frameworks (Li et al., 2022).

1.1. Definition of hybrid microgrids

Energy optimization involves reducing energy consumption while maintaining or improving production output. One potential solution for reducing energy consumption in the cement industry is using hybrid microgrids. A hybrid microgrid is a system that combines multiple sources of energy, including renewable energy sources such as solar and wind, with traditional energy sources such as diesel generators. Hybrid microgrids can be optimized to reduce energy costs, improve energy reliability, and reduce carbon emissions.

The potential benefits of using hybrid microgrids in the cement industry are significant. By combining multiple sources of energy, hybrid microgrids can reduce reliance on traditional fossil fuels, which can be costly and contribute to greenhouse gas emissions. Additionally, hybrid microgrids can help reduce the energy demand from the grid and provide backup power in case of grid outages, improving energy reliability.

1.2. Importance of hybrid microgrids in the contemporary applications

Hybrid microgrids have become indispensable in contemporary applications, serving as a critical solution for various energy challenges. These microgrids offer a range of benefits that address pressing concerns in the energy sector. Firstly, hybrid microgrids enhance energy resilience by integrating multiple energy sources and storage systems. This diversity ensures a reliable power supply even in the event of disruptions or failures in one energy source. The seamless switching capability of hybrid microgrids promotes uninterrupted operations, making them ideal for critical facilities, remote areas, and industries with stringent power requirements.

Secondly, hybrid microgrids play a vital role in the integration of renewable energy sources, such as solar and wind, into the power grid. By combining renewable energy with conventional sources and energy storage systems, hybrid microgrids effectively manage the intermittent nature of renewables and provide a stable power supply. This integration reduces greenhouse gas emissions, promotes environmental sustainability, and supports the transition towards a cleaner and more sustainable energy system. Furthermore, hybrid microgrids enable cost optimization by leveraging the advantages of different energy sources. The utilization of renewable energy sources with zero fuel costs, coupled with efficient load management and storage systems, helps optimize the overall cost of electricity generation and consumption.

1.3. Purpose of the chapter

The purpose of this chapter is to provide an in-depth exploration of hybrid microgrids in the context of contemporary applications. It aims to define hybrid microgrids and highlight their importance in various sectors. The chapter seeks to examine the applications of hybrid microgrids in remote communities, military installations, industrial and commercial complexes, universities and campuses, disaster resilience, data centers, resorts and tourism facilities, and rural electrification. Additionally, it aims to analyze the energy demand in these applications and identify the factors influencing energy demand. The chapter further explores the energy sources used in hybrid microgrids, including renewable and traditional sources, and discusses the benefits and challenges associated with each. Mathematical modeling, objective parameters, and case studies are also presented to provide a comprehensive understanding of hybrid microgrids. The overarching purpose of this chapter is to enhance knowledge and understanding of hybrid microgrids and their potential to address energy needs in diverse applications.

1.4. Objectives

(1) To provide an in-depth exploration of hybrid microgrids, including their applications, energy demand analysis, and energy sources.
(2) To present mathematical modelling and objective parameters for evaluating the performance and environmental impact of hybrid microgrids.
(3) To examine case studies and provide guidance for designing and implementing efficient and sustainable hybrid microgrid systems.

1.5. Contribution

The contribution of this chapter is that it provides an up-to-date overview of hybrid microgrids, including their applications, energy demand analysis, energy sources, mathematical modelling, objective parameters, and case studies. It highlights the importance of balancing the energy mix for optimal performance and sustainability. The document also presents guidance for designing and implementing efficient and sustainable hybrid microgrid systems. By providing a detailed analysis of the subject area, this document can be a valuable resource for researchers, engineers, masters and PhD students, and policymakers interested in the development and evolution of hybrid microgrids. It contributes to the field by providing a comprehensive and practical guide for designing and implementing hybrid microgrid systems that are efficient, sustainable, and cost-effective.

1.6. Applications of hybrid microgrid

1.6.1. Remote communities, islands, and offshore platforms

Remote communities, islands, and offshore platforms often face challenges in accessing reliable and affordable electricity from the main grid. Hybrid microgrids can be deployed in these locations to provide a sustainable and independent source of power.

In (Hasan et al., 2023) research focuses on optimizing the sizing of microgrid components by incorporating green hydrogen technology. Notably, the green hydrogen production system and the microgrid under investigation are situated on different islands, addressing specific constraints. The study in (Anglani et al., 2023) discusses an innovative control strategy for integrating a wind turbine (WT) and an energy storage unit into an existing stand-alone microgrid that serves an oil and gas (O&G) rig.

1.6.2. Military installations

Military installations require secure and reliable energy systems to support their operations. Hybrid microgrids offer the advantage of enhancing energy security by reducing the reliance on vulnerable fuel supply chains. These microgrids can integrate renewable energy sources with backup generators and energy storage, allowing military bases to operate efficiently and sustainably even in remote or austere environments. In (Reich and Sanchez, 2023) researchers establish a framework for generating multiple hybrid microgrid designs and evaluating their resilience in diverse scenarios, such as changing weather conditions, fluctuating power demands, and extended timeframes that surpass initial planning estimates.

1.6.3. Industrial and commercial

Hybrid microgrids find valuable applications in industrial and commercial complexes where a continuous power supply is critical for operations. The researchers in (Rao et al., 2023) examines the potential applications of microgrids in railway transportation and formulates operational strategies for both independent and grid-tied microgrid systems. To assess the economic and environmental impacts of microgrids, a comprehensive analytical framework is established, utilizing system dynamics, life cycle analysis, and life cycle cost analysis. The study (Castellanos et al., 2023) presents a convex optimization model for managing energy within unbalanced microgrids (MGs) in a Local Energy Market (LEM). This model accounts for operational constraints, power quality requirements, and interactions.

1.6.4. Universities and campuses

Universities and large campuses often have high energy demands due to the presence of numerous buildings, facilities, and research centers. The study (Ali et al., 2023) presents a survey of campus prosumer microgrids, covering aspects such as energy management strategies, optimization methods, architectural approaches, storage options, and design tools. The survey encompasses a decade of previous research for a comprehensive analysis. The study in (Alshehri et al., 2023) explores the development of a hybrid microgrid at King Saud University campus in Riyadh to fulfil its electricity needs by harnessing solar and wind resources available on-site.

1.6.5. Disaster resilience

In disaster-prone areas, reliable electricity is crucial for emergency response efforts, communication, and critical infrastructure. Hybrid microgrids can play a vital role in disaster resilience by providing independent and resilient power systems. In (Zamani Gargari et al., 2023) the authors introduce a method for assessing resiliency in multi-energy microgrids. We propose the use of mobile energy providers to mitigate the adverse effects of natural disasters and enhance system resiliency.

1.6.6. Data centers

Data centers have significant energy requirements and often require uninterrupted power to maintain critical operations. Hybrid microgrids offer a reliable and sustainable solution by integrating renewable energy sources, such as solar and wind, with energy storage and backup systems. In (Faheem et al., 2019a) authors present a unique distributed routing protocol called CARP, designed for Smart Grid applications based on Cognitive Radio Sensor Networks (CRSNs). In (Faheem et al., 2021) paper discusses a dataset comprising measurements obtained through IMWSNs during monitoring and control events in the smart grid.

1.6.7. Resorts and tourism facilities

Resorts and tourism facilities often operate in remote or environmentally sensitive areas, where access to reliable electricity can be challenging. Hybrid microgrids can provide sustainable power solutions by harnessing renewable energy sources like solar and wind, along with energy storage and backup generators. The study presented in (Żóładek et al., 2023) is to conduct a thorough feasibility analysis of an innovative hybrid renewable energy system aimed at achieving high self-sufficiency. This system combines wind turbines and photovoltaic panels to meet the energy demands of a tourist resort in Agkistro, Greece. Excess energy is stored in a battery and a hydrogen tank, and a wood gasifier serves as a backup energy source. The researchers in (Faheem et al., 2019b) introduces an innovative approach for data collection in smart grids. It leverages software-defined mobile sinks (SDMSs) and wireless sensor networks (WSNs) via the Internet.

1.6.8. Rural electrification

In rural areas with limited access to the main grid, hybrid microgrids can facilitate rural electrification initiatives. By combining renewable energy sources with energy storage, these microgrids can bring electricity to remote communities, improving their quality of life, supporting economic development, and enabling access to education, healthcare, and communication services. The study in (Kamal et al., 2023) explores an improved electricity solution for rural areas in Uttarakhand, India. It involves the implementation of a self-sustaining microgrid that efficiently meets the region's energy needs at a cost-effective rate. The system emphasizes the optimal sizing and sensitivity analysis of the hybrid energy model.

2. Energy demand analysis

2.1. Analysis of energy demand in applications of hybrid microgrid

Remote communities, islands, and offshore platforms have varying energy demands depending on factors such as population size, economic activities, and infrastructure. While the energy demand in these areas is generally lower compared to urban areas, it still encompasses residential electricity needs, essential services like healthcare and education, and support for small-scale industries. The energy demand in military installations can be substantial due to operational requirements, including powering offices, training facilities, communication systems, and residential areas for personnel.

The energy demand of industrial and commercial complexes varies widely based on the size, sector, and nature of the activities conducted within them. These complexes require energy to power machinery, lighting, HVAC systems, and other equipment. Similarly, universities and campuses have diverse energy demands due to the presence of multiple buildings, research facilities, student accommodations, and recreational areas.

Energy demand in disaster resilience applications can surge during emergencies, encompassing the provision of power to emergency shelters, hospitals, communication systems, lighting, and critical infrastructure.

Data centers have high energy demands driven by the continuous operation of servers, cooling systems, and other IT infrastructure. Resorts and tourism facilities require energy for lighting, HVAC systems, water heating, kitchen equipment, recreational facilities, and guest accommodations.

Rural electrification projects entail energy demand for residential households, public lighting, small-scale agricultural activities, and basic services like schools and healthcare facilities.

2.2. *Factors affecting energy demand*

Several factors influence the energy demand of these applications. Population size and growth play a significant role, as larger populations in remote communities, military installations, and universities result in higher energy demands. Economic activity and industry type also impact energy requirements, with energy-intensive processes and larger-scale operations contributing to higher demand. Geographical factors, such as limited energy resources in remote areas or the need for robust infrastructure in disaster-prone regions, can further affect energy demand. Additionally, technological advancements, energy efficiency measures, and the integration of renewable energy sources can influence overall energy demand by promoting more sustainable and efficient energy consumption practices (Bashawyah and Qaisar, 2021; Qaisar et al., 2021). Understanding these factors is essential for developing tailored energy solutions to meet the specific demands of each application.

2.3. *Methods for energy demand analysis*

Energy demand analysis utilizes a range of methods to comprehensively assess and understand energy consumption patterns. One commonly employed approach is conducting energy audits, which involve on-site inspections, data collection, and analysis to identify areas of high energy consumption, inefficiencies, and potential energy-saving measures. Energy modelling and simulation techniques leverage computer-based models to simulate energy consumption and estimate future energy demand based on factors such as building characteristics, equipment efficiency, and weather data. These models provide valuable insights into energy demand and enable the evaluation of different energy management strategies.

Statistical analysis of historical energy consumption data is another effective method for energy demand analysis. By employing techniques such as regression analysis and time series analysis, statistical methods reveal trends, patterns, and correlations between energy demand and various influencing factors like weather conditions, occupancy levels, or production volumes. Surveys and questionnaires are valuable tools for collecting primary data on energy consumption and user behaviour, providing insights into energy use patterns, user preferences, and awareness of energy-saving practices. Scenario analysis, on the other hand, explores hypothetical scenarios to assess the potential impacts of policy changes, technological advancements, or shifts in user behaviour on future energy demand. These methods, in combination or individually, enable a comprehensive analysis of energy demand, supporting informed decision-making and the development of effective energy management strategies.

3. Energy sources for hybrid microgrids

Hybrid microgrids deployed in applications such as remote communities, islands, military installations, industrial complexes, and other contexts often utilize a combination of energy sources to meet their specific energy needs. These sources include renewable energy options like solar power and wind power, which provide clean and sustainable electricity generation. Diesel generators are commonly used as backup or primary power sources in situations where renewable energy generation is insufficient. Biomass energy derived from organic materials can also be utilized, while battery storage systems play a crucial role in storing excess renewable energy for later use and ensuring a stable power supply. Additionally, in areas with access to water resources, micro-hydro power can be integrated into hybrid microgrids. The combination and integration of these energy sources are tailored to the requirements and available resources of each application, aiming to provide a reliable, resilient, and environmentally friendly energy supply.

3.1. Renewable energy sources

Hybrid microgrids for the applications mentioned earlier incorporate a variety of renewable energy sources to meet their energy needs. Solar power, derived from photovoltaic panels, serves as a clean and abundant source of electricity. Wind power, harnessed through wind turbines, is utilized in locations with consistent wind patterns. Biomass energy, derived from organic materials, provides a renewable and carbon-neutral energy source. Additionally, micro-hydro power can be integrated where there is access to water resources. These renewable energy sources are combined with battery storage systems to store excess energy and ensure a stable power supply during periods of low renewable generation or high demand. By integrating multiple renewable energy sources, hybrid microgrids in remote communities, islands, military installations, industrial complexes, universities, and other applications can achieve reliable, sustainable, and environmentally friendly energy supply. Figure 1 shows the solar potential of Pakistan (Group, 2022).

The benefits of using renewable energy sources include reduced operating costs and lower carbon emissions. Additionally, renewable energy sources can provide power during peak energy demand, reducing the need for expensive peak energy pricing. However, renewable energy sources are dependent on weather conditions and may not always be available when needed. Additionally, the initial investment required for installing renewable energy sources can be high.

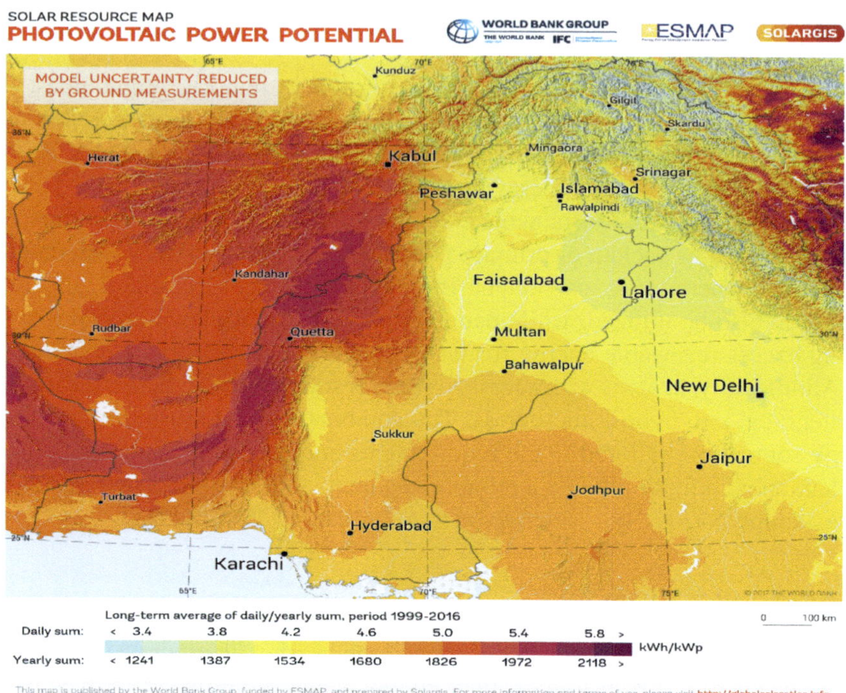

Figure 1. Solar potential map of Pakistan.

3.2. Traditional energy sources

In the context of the applications mentioned earlier, hybrid microgrids can utilize a combination of traditional energy sources to meet the diverse energy needs. These sources include grid connection, diesel generators, and possibly other conventional sources like natural gas or propane. Grid connection allows the hybrid microgrid to draw power from the main utility grid when available, ensuring a stable and reliable energy supply. Diesel generators serve as backup or primary power sources in areas with limited grid access or during periods of high demand. These traditional energy sources provide a reliable and readily available power supply, offering support during times of low renewable energy generation or peak demand. The integration of traditional energy sources alongside renewable sources in hybrid microgrids allows for a balanced and resilient energy system that can cater to the specific requirements of each application.

3.3. Benefits and challenges of using renewable and traditional energy sources

The utilization of both renewable and traditional energy sources in a hybrid microgrid offers a range of benefits. Figure 2 represents the overall microgrid en-

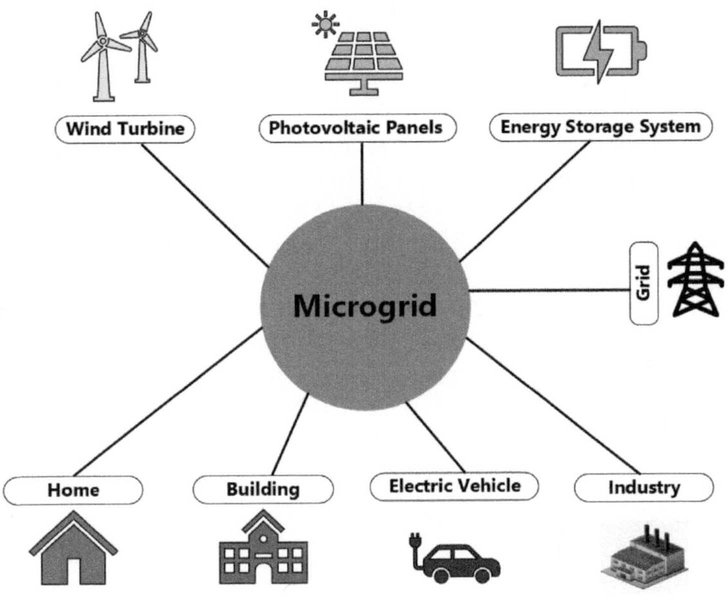

Figure 2. Overall microgrid environment.

vironment. Renewable energy sources such as solar and wind power provide significant environmental advantages, as they produce minimal greenhouse gas emissions and contribute to reducing reliance on fossil fuels. They also offer long-term cost savings by harnessing freely available and abundant resources. Additionally, renewable energy sources can enhance energy independence, particularly in remote areas or islands, by reducing dependence on external energy sources and increasing energy resilience. Integration with traditional energy sources, such as grid connection or diesel generators, provides backup power and ensures a reliable energy supply during periods of low renewable energy generation or high demand. This hybrid approach offers flexibility, stability, and the ability to balance energy generation and consumption effectively.

However, using both renewable and traditional energy sources in a hybrid microgrid also poses certain challenges. One challenge is the intermittent nature of renewable energy sources, such as solar and wind power, which depend on weather conditions and time of day. This intermittency requires effective energy management and storage solutions, such as battery storage systems, to ensure a continuous and reliable power supply. Additionally, the initial costs of installing renewable energy infrastructure can be relatively high, although they are often offset by long-term operational cost savings. Integration with traditional energy sources may introduce issues related to carbon emissions and environmental impacts associated with the use of fossil fuels. Balancing the energy mix and optimizing the system for efficiency and sustainability requires careful planning, design, and operational management of the hybrid microgrid.

4. Mathematical modelling of selected resources

4.1. Photovoltaic (PV)

A PV panel, also known as a solar panel or photovoltaic module, is a device that converts sunlight into electricity. It is made up of multiple photovoltaic cells connected, which work together to generate a usable electrical current. The cells are typically made of silicon and are arranged in a specific pattern to maximize their energy production. When sunlight strikes the surface of the cells, it excites the electrons within them, causing them to flow in a specific direction and generate a direct current (DC) electricity. This DC electricity can then be converted into alternating current (AC) electricity using an inverter, which makes it suitable for powering homes, businesses, and other types of electrical loads. PV panels are a key component of solar power systems and are widely used in residential, commercial, and utility-scale applications around the world (Abdel Gawad et al., 2022; Ammach and Qaisar, 2022).

Equation (Icaza-Alvarez et al., 2022) calculates the module's output power under normal working conditions.

$$P_{pv} = f_{pv} \times Y_{pv} \times \frac{I_T}{I_S} \tag{1}$$

The term P_{pv} represents the rated power output of photovoltaic (PV) panels, expressed in kilowatts (kW). Meanwhile, I_T refers to the total amount of solar radiation that falls on the panels, measured in kilowatt-hours per square meter (kWh/m2). The value of I_S, which is equal to 1000 watts per square meter (W/m2), represents the standard intensity of solar radiation under ideal conditions. Lastly, f_{pv} denotes the reduction factor that accounts for energy losses resulting from factors like long wiring distances and splices in the PV system.

4.2. Diesel generator

A diesel generator is a type of internal combustion engine that uses diesel fuel to generate electrical power. It consists of a diesel engine and an electric generator, which work together to convert diesel fuel into electricity. The diesel engine is responsible for converting the chemical energy in the diesel fuel into mechanical energy, which then drives the electric generator to produce electricity. Diesel generators are commonly used as backup power sources in situations where grid power is unavailable or unreliable, such as in remote areas, construction sites, or during power outages.

The connection depicted in Equation 2 from (Zieba Falama et al., 2022) establishes a connection between a diesel generator's output and rated power.

$$PGD = \eta\, diesel \times NDG \times PGD, N(2) \tag{2}$$

where NDG is total number of identical diesel generators, PDG is the combined output power of the generators, and η is the efficiency of the generator.

The planned hybrid system's estimated CO2 emissions were calculated using the Equation 3 from (Shezan et al., 2022):

$$tCO_2 = 3.667 \times m_f \times Hv_f \times CEF_f \times x_c \qquad (3)$$

The quantity of fuel used is denoted by m_f, while the total CO2 emissions are represented by tCO_2. Additionally, the abbreviations Hv_f, CEF_f, and x_c refer to Tons of Carbon Emitted per Terajoule, the percentage of oxidized carbon, and Heating Value of Fuel in Megajoules per liter, respectively. It is worth noting that one gram of carbon is present in 3.667 grams of CO2, which is an important factor to consider in these calculations.

4.3. Fuel cell

A fuel cell is an electrochemical device that converts the chemical energy of a fuel into electrical energy. Unlike conventional power sources, such as combustion engines, fuel cells produce electricity without burning the fuel. This means that they can operate at much higher efficiencies, with lower emissions, and with less noise than traditional power sources.

The basic components of a fuel cell include an anode, a cathode, and an electrolyte. The fuel is introduced to the anode, and the oxidant (usually oxygen from the air) is introduced to the cathode. The fuel is then oxidized at the anode, releasing electrons that flow through an external circuit to the cathode, producing electricity. At the cathode, the electrons combine with the oxidant and any remaining fuel to form water and other by products.

The net output of fuel cell is given by Equation (4) from (Wishart et al., 2006).

$$W_{netoutput} = Ixv_{cell} \times n_{cell} \times \eta_{net} \qquad (4)$$

where n_{cell} is the number of cells in the stack. η_{net} is the efficiency of fuel cell.

4.4. Electrolyzer

An electrolyzer is an electrochemical device that uses electricity to split water molecules into hydrogen and oxygen gases through a process called electrolysis. The basic components of an electrolyzer include an anode, a cathode, and an electrolyte, which is usually a solution of potassium hydroxide.

During electrolysis, an electric current is applied to the anode and cathode, which are separated by the electrolyte. This causes water molecules to break apart, with the positively charged hydrogen ions (protons) migrating to the cathode and the negatively charged oxygen ions migrating to the anode. At the cathode, the protons combine with electrons from the electric current to form hydrogen gas, while at the anode, the oxygen ions combine with water molecules to form oxygen gas and positively charged hydrogen ions.

The hydrogen gas produced by an electrolyzer can be used as a fuel for vehicles or to generate electricity in a fuel cell. It can also be stored for later use, either as a gas or by compressing it into a liquid. Electrolyzers can be powered by renewable energy sources like solar or wind, making them an important tool for producing carbon-free hydrogen.

While electrolyzers are still relatively expensive and not yet widely deployed, they have the potential to play a significant role in a future low-carbon energy system, particularly as renewable electricity becomes more abundant and affordable.

Equation 5 provides the necessary power, P_r, for the electrolyzer to function, which is determined by U_c, the voltage supplied to the electrolyzer, N_c, the number of cells connected in series, and I_{el}, the current flowing through the outer circuit. According to Faraday's law, the amount of hydrogen produced by the electrolyzer is directly proportional to the rate of electron exchange at the terminals, which is equivalent to the electrical current flowing in the outer circuit (Ghennou et al., 2022).

$$P_r = U_c \times N_c \times I_{el} \tag{5}$$

4.5. Hydrogen tank

A hydrogen tank is a specialized container used to store hydrogen gas under high pressure. Due to its low density, hydrogen gas must be compressed to a high pressure in order to store a sufficient amount of fuel in a reasonably sized container. Hydrogen tanks are typically made of composite materials or high-strength metals, such as carbon fibre or aluminium, and are designed to withstand the high pressure of the stored hydrogen.

Hydrogen tanks are a critical component of hydrogen fuel cell vehicles, where they are used to store the hydrogen fuel that powers the fuel cell. The size and capacity of the hydrogen tank can vary depending on the vehicle and the intended use, but they typically store between 3 and 7 kilograms of hydrogen gas at pressures ranging from 350 to 700 bar.

One of the challenges of using hydrogen tanks is the potential for hydrogen embrittlement, a phenomenon in which the high-pressure hydrogen gas can weaken or damage the material of the tank over time. To mitigate this risk, tanks are often designed with special linings or coatings to protect against hydrogen embrittlement.

Overall, hydrogen tanks are a critical component in the storage and transportation of hydrogen fuel, and continued advancements in their design and manufacturing will play an important role in the widespread adoption of hydrogen as a clean and sustainable energy source.

4.6. Energy storage systems for hybrid microgrids

4.6.1. Types of energy storage systems

Energy storage systems are an essential component of hybrid microgrids as they provide a reliable and consistent source of energy when the primary sources, such as solar or wind, are not available. There are various types of energy storage systems that can be used in hybrid microgrids, including:

(1) Battery energy storage systems (BESS): BESS uses batteries to store and release energy. Lithium-ion batteries are the most used batteries in BESS due to their high energy density, efficiency, and low maintenance requirements (Khan et al., 2022; Qaisar, 2020).

(2) Flywheel energy storage systems: Flywheels store energy in the form of kinetic energy by rotating a rotor at high speeds. The energy can be released when needed by converting the kinetic energy into electrical energy.

4.6.2. Benefits of energy storage systems in hybrid microgrids

Energy storage systems provide several benefits in hybrid microgrids, including:

(1) Improved reliability: Energy storage systems can provide backup power during power outages, ensuring uninterrupted power supply to critical equipment.

(2) Cost savings: Energy storage systems can help reduce energy costs by storing excess energy generated during low demand periods and releasing it during high demand periods, reducing the need to purchase energy from the grid.

(3) Carbon emission reduction: Energy storage systems can help reduce carbon emissions by allowing the use of renewable energy sources such as solar and wind power even when they are not available.

(4) Increased efficiency: Energy storage systems can improve the efficiency of hybrid microgrids by balancing the energy supply and demand, reducing energy wastage, and optimizing the use of energy resources.

In conclusion, energy storage systems are critical components of hybrid microgrids in the cement industry. They provide several benefits, including improved reliability, cost savings, carbon emission reduction, and increased efficiency. The selection of the appropriate energy storage system depends on various factors such as the plant's energy demand, location, and cost-effectiveness.

4.6.3. Battery energy storage systems (BESS)

The BESS stands for Battery Energy Storage System, which is a type of energy storage system that uses batteries to store energy for later use. BESSs are becoming increasingly popular as renewable energy sources like solar, and wind become more common, as they allow excess energy to be stored when it is available and used when it is needed (Mian Qaisar, 2020).

The basic components of a BESS include the batteries themselves, a control system, and inverters to convert the DC power from the batteries into AC power for use in homes, businesses, or the grid. BESSs can be used for a variety of applications, from providing backup power during outages to storing energy from renewable sources for use during times of high demand.

One of the key advantages of BESSs is their ability to provide fast response times, making them ideal for use in frequency regulation and other grid stabilization services. They can also help reduce the need for expensive infrastructure upgrades by providing localized energy storage to offset peak demand.

While BESSs are still relatively expensive, their costs are expected to continue to decline as battery technology improves and production scales up. As the need for energy storage grows, BESSs are likely to play an increasingly important role in a future low-carbon energy system.

From Equation 6 as in (Omotoso et al., 2022),

$$SOC\left(t\right) = SOC\left(t-1\right)\left(1-\sigma\right) + \left(P_{GA}(t) - \frac{P_L(t)}{\eta_{inv}}\right)\eta_{battery} \qquad (6)$$

The load demand is represented by $P_L(t)$, while the battery's state of charge is indicated by $SOC(t)$. The ceil function classifies the expression as part of a group of expressions that are near or equivalent to the total. In hybrid energy systems, battery storage not only provides storage but also helps maintain a balance between the electricity supply and demand. The system measures energy output, consumption, and charging status over time.

4.7. Converter

Converter is that can function as both a rectifier and an inverter. During periods of low solar and wind resources, such as at night or on cloudy days, the converter operates solely in inverter mode. Conversely, when there is enough renewable energy available to charge the battery storage system, the converter operates solely in rectifier mode (Ayad et al., 2023; Ilahi et al., 2023).

Equation 7 in Prakash and Dhal (2021) expresses the maximal capacity of the power converter to convert DC to AC, which depends on the choice of inverter and its efficiency (Pl, s(t))

$$P_{l,s}\left(t\right) = P_{input}\left(t\right) * \eta_{conv} \qquad (7)$$

where $P_{l,s}\left(t\right)$ denotes the converter's input power and eta_{conv} the efficiency of the converter.

5. Objectives parameters

To find the most optimal solution, net present cost (NPC), levelized cost of electricity (LCOE) and greenhouse gases (GHG) emissions are considered as an objective. By altering the component's values, the output may be modified by doing so, authors can improve the system's NPC, LCOE and GHG emissions.

5.1. Net present cost

The total ongoing costs of the system over its useful life, minus the recovery value during that period, are equivalent to the Net Present Cost (NPC) incurred by the framework. According to reference (Shezan et al., 2022), the costs considered for net present cost calculation include capital expenses, replacement costs, operational expenses, and maintenance costs as shown in Equation 8. Homer Pro software is used to calculate the NPC for each component of the installed system.
The formula below is used to determine the total NPC:

$$C_{NPC} = \frac{C_{ann.tot}}{CRF\,(i.R_{proj})} \tag{8}$$

Here, $C_{ann.tot}$ is the Annualized cost, i the Interest rate (Annual), R_{proj} the project lifetime, and $CRF(.)$ the Capital recovery factor.

5.2. Levelized cost of electricity

A typical cost per KWh of power is delivered by the predetermined shaped framework. To determine the optimal COE for a standalone system, HOMER uses the Equation 9 from (Shezan et al., 2022).

$$LCOE = \frac{C_{ann.tot}}{E_{prim} + E_{def} + E_{grid.sales}}, \tag{9}$$

where E_{prim} is the total primary $C_{ann.tot}$ is the annualized total cost, E_{def} is the total deferrable load, and $E_{grid.sales}$, sales are the amount of energy sold to the grid (per year).

5.3. GHG emissions

Energy generation results in the release of harmful gas emissions, which depend on the type of energy source used. The amount of carbon dioxide emitted per kWh varies depending on the energy source used, resulting in a fluctuation of emissions levels from year to year. Additionally, every kWh produced results in the emission of 1.34 g of nitrogen oxides and 2.74 g of carbon dioxide. However, none of these harmful gases, including nitrogen oxides (NO), sulphur dioxide (SO2), carbon monoxide (CO), unburned hydrocarbons (UHCs), or carbon dioxide (CO2), are present in the renewable hybrid HEM-1 or HEM-3.

6. Case studies

6.1. Case study of cement industry

The researcher in (Basheer et al., 2022) studied the hybrid energy models (HEMs) in cement industry of Pakistan. The authors selected 4 different HEMs and 5 cement plants to get the objective parameter NPC, LCOE and GHG emissions.

For this study, five cement plants were under consideration:

(1) Cement Plant-1: Askari Cement Plant, Wah (ACPW).
(2) Cement Plant-2: Bestway Cement Plant, Kalar Kahar (BCPKK).
(3) Cement Plant-3: Bestway Cement Plant, Farooqia (BCPF).
(4) Cement Plant-4: Bestway Cement Plant, Hattar (BCPH).
(5) Cement Plant-5: DG Cement Plant, Chakwal (DGCPC).

Figure 3 shows a list of the four types of HEMs that are created for this hybrid renewable system.

(1) HEM-1: As shown in Figure 3(a), it will have a PV, hydrogen tank, converter, electrolyzer, and fuel cell.
(2) HEM-2: As shown in Figure 3(b), it will only have a diesel generator.
(3) HEM-3: As shown in Figure 3(c), it will have a PV, converter, and battery framework.
(4) HEM-4: As shown in Figure 3(d), it will have a diesel generator, PV, and Converter.

6.1.1. Cost analysis

The results of HEMs in cement industry of Pakistan in Tables 1–5 presents the initial cost, operating cost, LCOE, and NPC of all HEM plant costs. Figure 4 displays the consolidated results of NPC of each model for all plants. ACPW NPC for HEM-1 and HEM-2 is US$2630M and US$575M, respectively, while HEM-3 and HEM-4 have corresponding NPCs of US$4970M and US$540M. The NPC for BCPKK is US$2890M for HEM-1, and for HEM-2, HEM-3, and HEM-4, it is US$1080M, US$9380M, and US$1010M, respectively. The NPC for BCPF is US$2940M for HEM-1, and for HEM-2, HEM-3, and HEM-4, it is US$1180M, US$1050M, and US$1100M, respectively. The NPC for BCPH is US$2630M for HEM-1, and for HEM-2, HEM-3, and HEM-4, it is US$575M, US$4970M, and US$540M, respectively. The NPC for DGCPC is US$2830M for HEM-1, US$187M for HEM-2, US$8,480M for HEM-3, and US$923M for HEM-4. Figure 5 illustrates the LCOE of cement plants. The LCOE for ACPW from HEM-1 and HEM-2 is US$1.22 and US$0.266, respectively, while HEM-3 and HEM-4

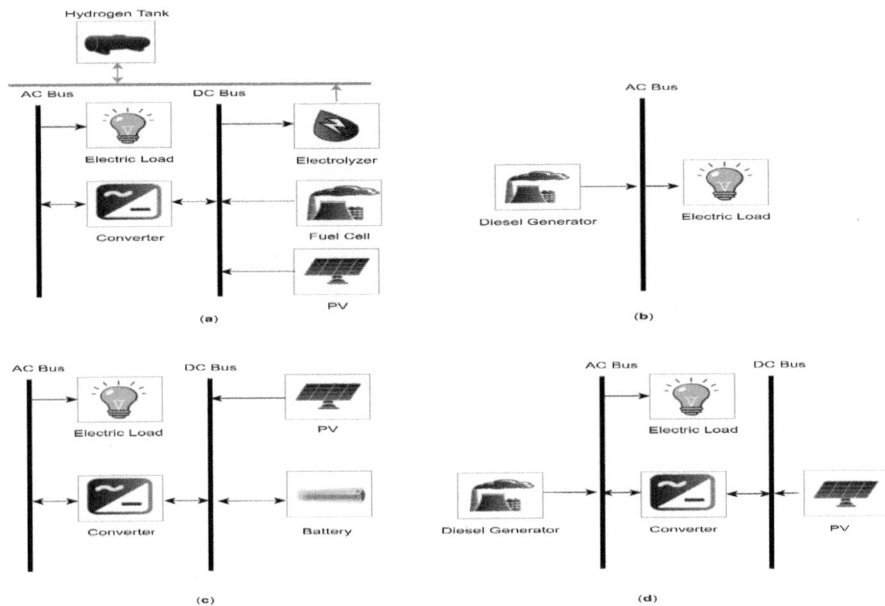

Figure 3. Schematic diagram of HEMs: (a) HEM-1; (b) HEM-2; (c) HEM-3; (d) HEM-4.

have corresponding LCOEs of US$2.30 and US$0.249. The LCOE for BCPKK is US$0.706 for HEM-1, and for HEM-2, HEM-3, and HEM-4, it is US$0.265M, US$2.30M, and US$0.248, respectively. The LCOE for BCPF is US$0.660 for HEM-1, and for HEM-2, HEM-3, and HEM-4, it is US$0.264, US$2.20, and US$0.248, respectively. The LCOE for BCPH is US$1.22 for HEM-1, and for HEM-2, HEM-3, and HEM-4, it is US$0.266, US$2.30, and US$0.249, respectively. The LCOE for DGCPC is US$0.706 for HEM-1, US$0.265 for HEM-2, US$2.28 for HEM-3, and US$0.248 for HEM-4.

Table 1. HEM-1, 2, 3, and 4. Comparison of ACPW's NPC, initial cost, COE, and O&M.

Cost type	HEM-1	HEM-2	HEM-3	HEM-4
NPC	US$2630M	US$575M	US$4970M	US$540M
Initial cost	US$634M	US$14.8M	US$3060M	US$95.8M
COE/kWh	US$1.22	US$0.266	US$2.30	US$0.249
O&M	US$146M/yr	US$40.8M/yr	US$4,805M/yr	US$32.4M/yr

Table 2. HEM-1, 2, 3, and 4. Comparison of BCPKK's NPC, initial cost, COE, and O&M.

Cost type	HEM-1	HEM-2	HEM-3	HEM-4
NPC	US$2890M	US$1080M	US$9380M	US$1010M
Initial cost	US$743M	US$27.6M	US$5780M	US$175M
COE/kWh	US$0.706	US$0.265	US$2.30	US$0.248
O&M	US$156M/yr	US$76.8M/yr	US$263M/yr	US$61.1M/yr

Table 3. HEM-1, 2, 3, and 4. Comparison of BCPF's NPC, initial cost, COE, and O&M.

Cost type	HEM-1	HEM-2	HEM-3	HEM-4
NPC	US$2940M	US$1180M	US$1050M	US$1100M
Initial cost	US$764M	US$30M	US$6280M	US$190M
COE/kWh	US$0.660	US$0.264	US$2.20	US$0.248
O&M	US$158M/yr	US$83.5M/yr	US$287M/yr	US$66.6M/yr

Table 4. HEM-1, 2, 3, and 4. Comparison of BCPH's NPC, initial cost, COE, and O&M.

Cost type	HEM-1	HEM-2	HEM-3	HEM-4
NPC	US$2630M	US$575M	US$4970M	US$540M
Initial cost	US$634M	US$14.8M	US$3060M	US$94.5M
COE/kWh	US$1.22	US$0.266	US$2.30	US$0.249
O&M	US$146M/yr	US$40.8M/yr	US$139M/yr	US$32.5M/yr

Table 5. HEM-1, 2, 3, and 4. Comparison of DGCPC's NPC, initial cost, COE, and O&M.

Cost type	HEM-1	HEM-2	HEM-3	HEM-4
NPC	US$2830M	US$987M	US$8480M	US$923M
Initial cost	US$720M	US$25.2M	US$5230M	US$159M
COE/kWh	US$0.760	US$0.265	US$2.28	US$0.248
O&M	US$154M/yr	US$70M/yr	US$236M/yr	US$55.6M/yr

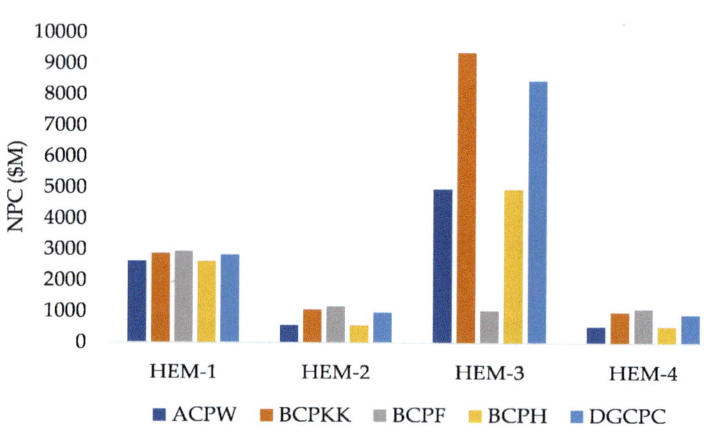

Figure 4. Comparison of the NPC for cement plants.

6.1.2. GHG emissions

The sustainable HEMs used in HEM-1 and HEM-2 do not harm the environment through the emission of harmful gases. While the generator in the hybrid HEM-2 and HEM-4 does produce harmful gases, it has been limited to producing only the bare minimum of energy during crises to minimize environmental damage. Similarly, the output of the fuel cell in HEM-1 has been restricted

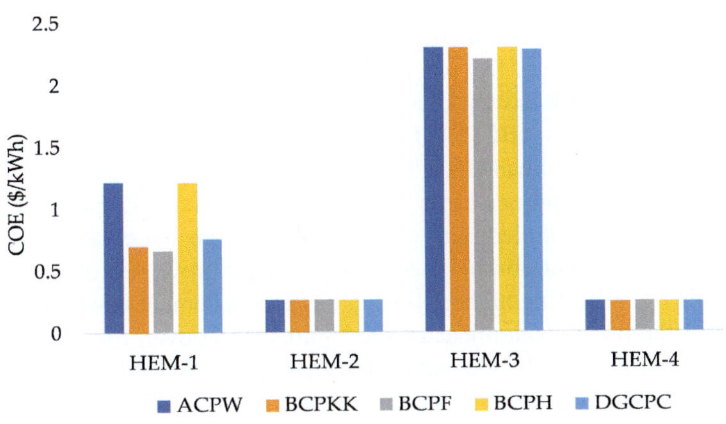

Figure 5. Comparison of the LCOE for cement plants.

to reduce the production of harmful gases, as fuel cells emit no carbon dioxide. Neither HEM-1 nor HEM-2 produces any greenhouse gas emissions. However, HEM-2 does produce significant emissions of 111594.85 tons annually for ACPW, 210469.811 tons annually for BCPKK, 229008 tons annually for BCPF, 111594.847 tons annually for BCPH, and 191930 tons annually for DGCPC, as indicated in Figure 6.

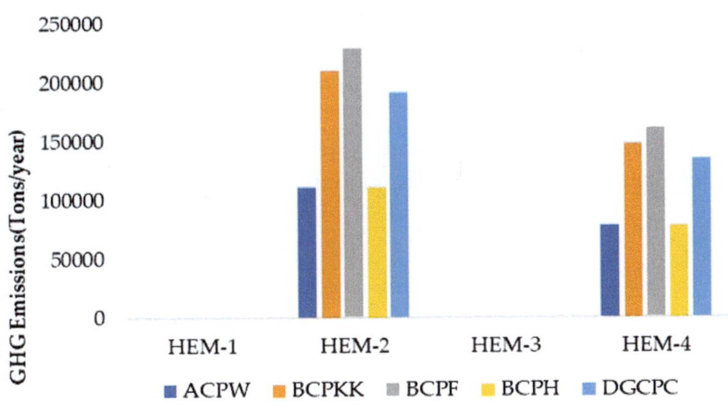

Figure 6. Comparisons of HEMs with respect to GHG emissions for cement plants.

6.1.3. Discussion

The study conducted a comprehensive assessment of hybrid energy models (HEMs) within Pakistan's cement industry, examining four distinct configurations: HEM-1 incorporating PV panels, hydrogen storage, converter, electrolyzer, and fuel cell; HEM-2 with a diesel generator; HEM-3 featuring PV panels, converter, and battery; and HEM-4 combining a diesel generator, PV panels, and converter. Inclusion of five prominent cement plants, namely Askari Cement Plant, Wah (ACPW); Bestway Cement Plant, Kalar Kahar (BCPKK); Bestway Cement Plant, Farooqia (BCPF); Bestway Cement Plant, Hattar (BCPH); and DG Cement Plant, Chakwal (DGCPC), enabled a comprehensive analysis. The study meticulously examined costs, encompassing initial and operating costs, as well as levelized cost of electricity (LCOE) and net present cost (NPC) for each HEM and cement plant combination. The outcomes revealed distinct patterns, such as HEM-1 indicating higher NPC values across cement plants, while HEM-2 displayed the most economical LCOE figures. Moreover, the research underscored the environmental dimension, designating HEM-1 and HEM-2 as environmentally benign due to minimal greenhouse gas emissions. Notably, HEM-4 demonstrated emissions of 111,594.85 to 191,930 tons/year across different cement plants. These findings collectively provide a comprehensive overview of the economic, emissions, and efficacy aspects of various hybrid energy models in Pakistan's cement industry.

6.2. Case study of Gwadar

In (Ali et al., 2022) the authors developed renewable energy system where three distinct models were created, encompassing photovoltaics, wind turbines, batteries, grid connections, and the specified capacity mentioned earlier. Schematic representations of the three model designs can be seen in Figures 7 to 9. The proposed systems were located at coordinates 4867+5Q Gwadar, Pakistan (25J6.6" N, 62J18.9" E), and were composed of solar panels, wind turbines, battery banks, and grid connectivity, reflecting the recommended renewable energy system scenario.

The Cost Analysis of the proposed system is shown in Figures 10–12. The economic evaluation of three models, namely Model 1, Model 2, and Model 3,

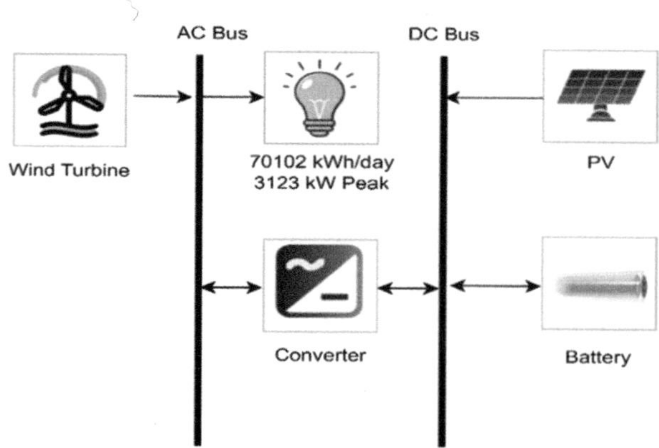

Figure 7. Model 1.

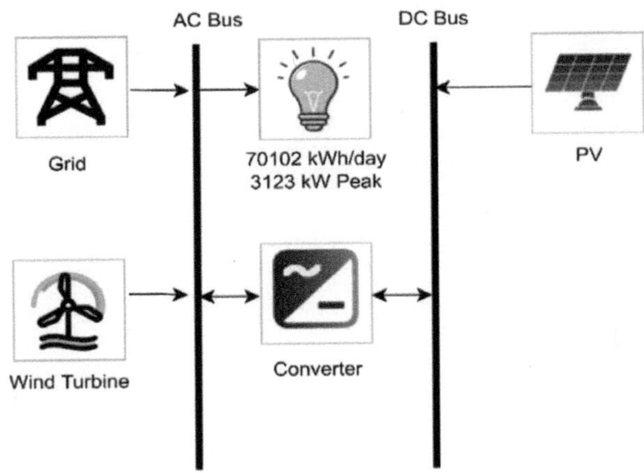

Figure 8. Model 2.

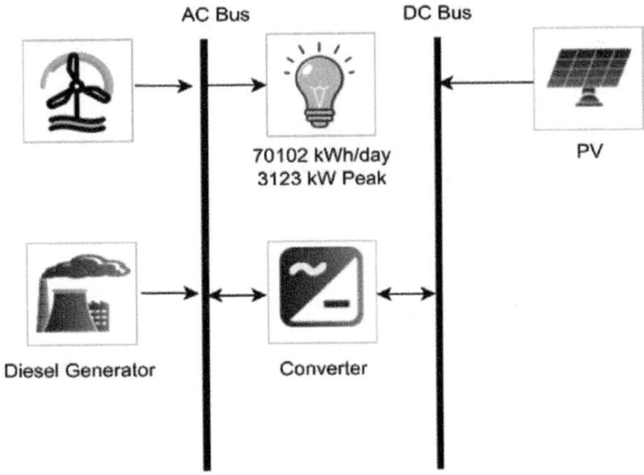

Figure 9. Model 3.

resulted in net present costs (NPC) of USD 166 million, USD 28.2 million, and USD 76.4 million, respectively. Model 1 incurs a higher cost primarily due to the inclusion of battery storage as the sole backup power source. Conversely, Model 3 incorporates a diesel generator to offset fuel expenses, contributing to its elevated NPC. Considering these factors, Model 2 emerges as the preferred choice. Notably, Model 2 demonstrates the lowest NPC among the three models, consequently yielding the lowest levelized cost of electricity (LCOE). Specifically, Model 2 exhibits an LCOE of USD 0.0347/kWh, while Model 1 and Model 3 exhibit LCOE values of USD 0.401/kWh and USD 0.184/kWh, respectively. The LCOE of Model 1 is even lower than Pakistan's current tariff rate (pap). Additionally, the salvage costs for Model 1, Model 2, and Model 3 are USD 42,167,420, USD 1,405,500, and USD 785,865, respectively.

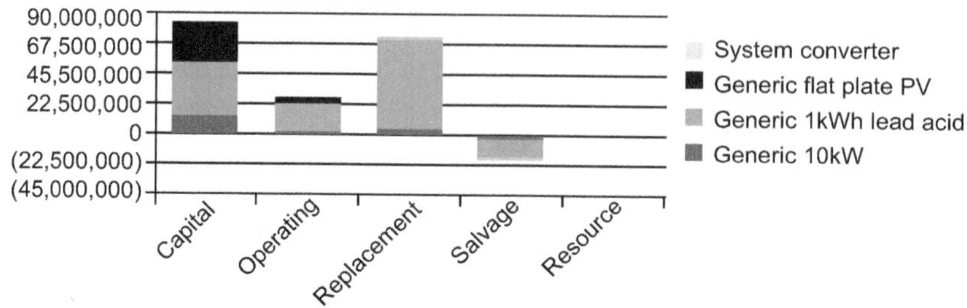

Figure 10. Cost graph for Model 1.

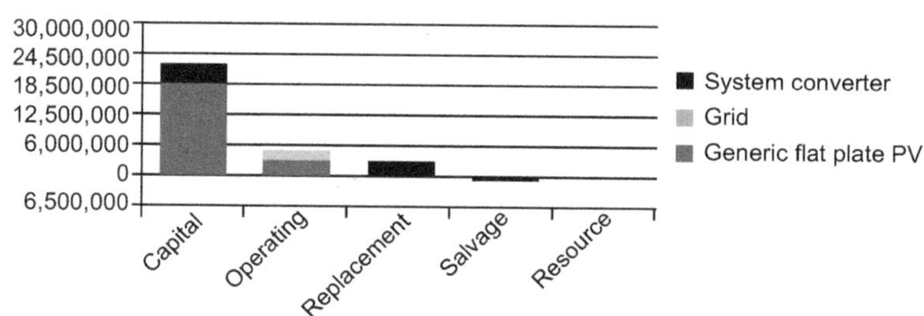

Figure 11. Cost graph for Model 2.

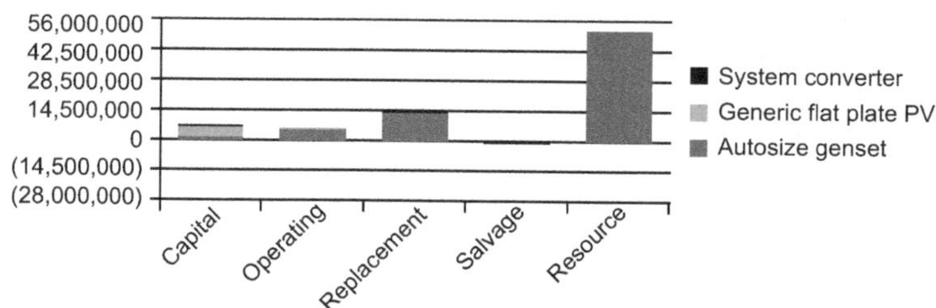

Figure 12. Cost graph for Model 3.

6.2.1. Sensitivity analysis

A sensitivity analysis was conducted to evaluate the impact of certain parameters on the economic and technical performance of the proposed system. Two examples of these parameters are the discount rate and wind turbine hub height. Figure 13 illustrates the significant effect of the discount rate on the economic viability of the three models. The net present cost (NPC) decreases as the discount rate increases. The discounting process is crucial in such analyses as it allows decision-makers and investors to understand the long-term consequences and costs associated with the project. By adjusting the difference between present and future values, stakeholders can consistently compare the benefits and costs of the proposed policies. In this case, both the NPC and the levelized cost of electricity (LCOE) decrease as the discount rate increases, indicating a direct relationship between the two variables.

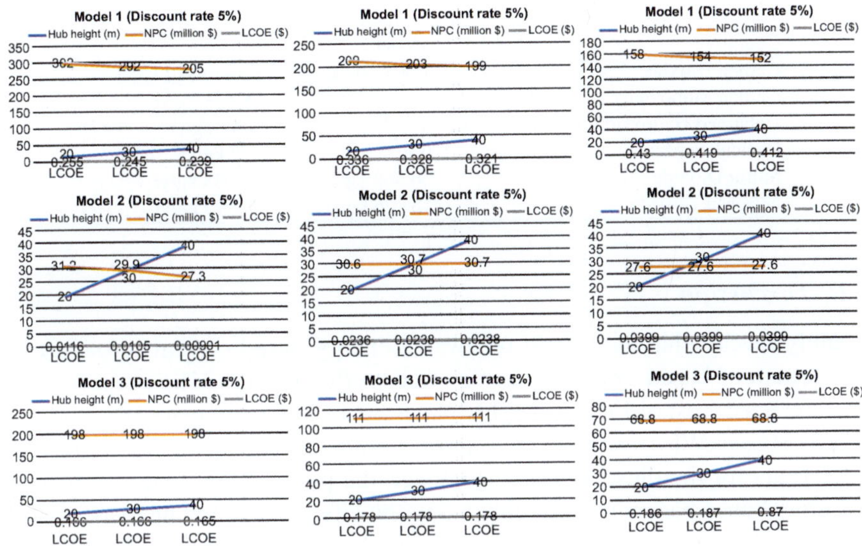

Figure 13. Sensitivity analysis.

Additionally, the impact of the wind turbine hub height on the NPC and LCOE is evident in the table. An increase in hub height results in a decrease in both the NPC and LCOE. This can be attributed to the fact that higher hub heights offer more stable wind speeds, leading to a consistent generation of electricity. The stability in electricity production contributes to the reduction in both the NPC and LCOE.

6.2.2. Environmental impact of model

In the present day, renewable energy sources (RES) have gained significant recognition as eco-friendly options for electricity generation, contributing to the decarbonization of the energy sector. RES, while not producing carbon dioxide emissions, still warrant careful consideration due to potential environmental impacts. Thus, selecting appropriate RES for power generation at any given plant is crucial. Thorough analysis reveals that if RES selection is not done meticulously, it can result in harmful environmental consequences. Greenhouse gas emissions comparison across three models is presented in Table 6. Model 1 exclusively employs RES, reflecting a renewable fraction of 100%. Conversely, Model 2 relies on grid-purchased electricity, resulting in a lower renewable energy fraction of 73.3%. Model 3, relying solely on diesel generators for backup power, exhibits a notably low renewable fraction.

Table 6. GHG emissions.

Quantity	Model 1	Model 2	Model 3
Carbon Dioxide	0	8,501,900 kg/yr	12,510,312 kg/yr
Carbon Monoxide	0	0	78,858 kg/yr
Unburned Hydrocarbons	0	0	3441 kg/yr
Particulate Matter	0	0	478 kg/yr
Sulfur Dioxide	0	36,860 kg/yr	30,635 kg/yr
Nitrogen Dioxide	0	18,026 kg/yr	74,079 kg/yr

6.2.3. Discussion

The study reveals that the hybrid renewable energy-based system surpasses the diesel generator-based system in terms of cost-effectiveness. The proposed system's Levelized Cost of Electricity (LCOE) is \$0.12/kWh, compared to the diesel system's \$0.28/kWh, resulting in a shorter payback period of 5.5 years versus 7.5 years for the diesel option. Moreover, the proposed system significantly curbs greenhouse gas emissions, enhancing the city's environmental sustainability. Carbon dioxide emissions are projected to decrease by up to 12.5 million kg annually, with sulfur dioxide and nitrogen dioxide emissions dropping by up to 36,860 kg and 74,079 kg per year, respectively. In essence, the study underscores that the hybrid renewable energy-based system presents a viable, economical solution to Gwadar city's electricity deficit. With its reliability, sustainability, and emission reduction benefits, it holds the potential to tackle energy challenges while enhancing the city's environmental well-being.

6.3. Case study of University Campus

The study in (Awan et al., 2022) delves into the comprehensive discussion of the design, development, and optimization of microgrids specifically tailored for the MUST University Mirpur campus. The microgrid design, which encompasses the utilization of all available resources at the university, is presented in Figure 14, while the proposed optimal design is depicted in Figure 15. Initially, a thorough modelling of the available resources at MUST University was conducted. In this model, the DC voltage sources, namely the SG200M5 solar photovoltaic panels and the 1 kWh L battery bank, were connected to the DC bus. On the other hand, the AC voltage sources, the GEN 100 diesel generator, and the main grid, were connected to the AC bus. To facilitate the bidirectional flow of electrical energy as needed, a bidirectional power converter was incorporated to link the DC and AC bus bars. Currently, the load is connected to the AC bus bar, but it can be modified to accommodate a DC load with a dedicated DC bus bar.

Through simulations based on relevant parameters and data, the optimal design proposed by HOMER demonstrated that the microgrid system could operate entirely independent of the grid throughout the entire year. As a result, the grid was excluded from the proposed optimal microgrid design for the MUST

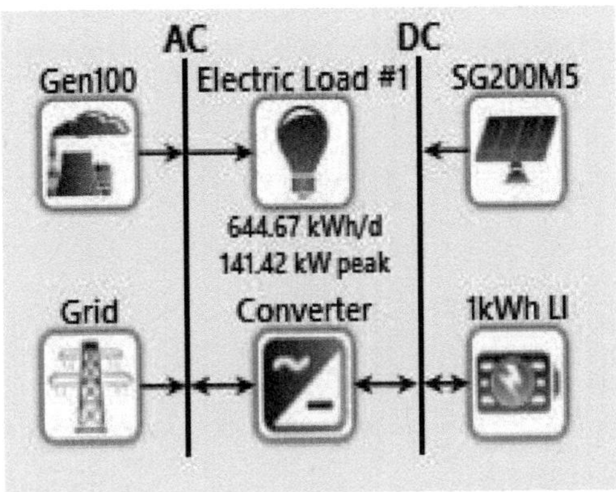

Figure 14. Microgrid structure with available resources at MUST site.

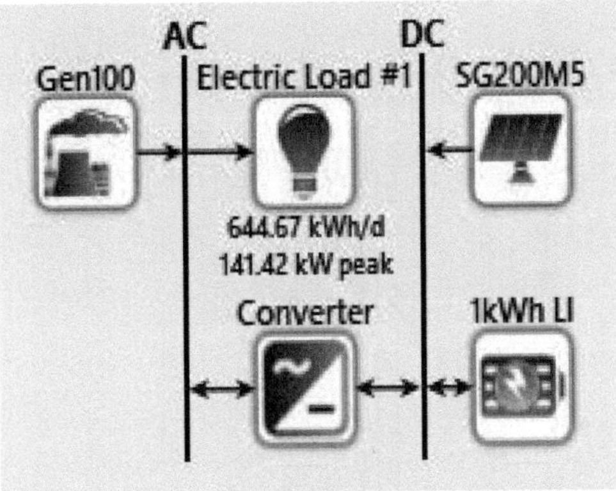

Figure 15. Proposed microgrid for the MUST site.

University Mirpur campus, as illustrated in Figure 13. This indicates the self-sufficiency and autonomy of the microgrid system, allowing it to rely solely on the available resources and efficiently manage the electrical energy requirements of the university without external grid support.

To achieve the most efficient outcomes, the upper and lower limits for each energy source were determined through the utilization of the "HOMER Optimization" feature. By employing this approach, the HOMER software conducted numerous simulations, leading to the identification of an optimized microgrid design. The results of this optimization process are presented in Table 7. The proposed microgrid designs for the Mirpur campus of MUST were generated based on five key performance criteria, namely net operating cost, cost of energy, operating cost, capital cost, and renewable fraction.

Table 8 presents a comprehensive economic comparison of the various proposed optimal scenarios or categories of a microgrid. The analysis reveals that

Table 7. Optimized design of microgrid proposed by HOMER.

Category	PV(KW)	Genset (KW)	Battery (KWh)	Grid (KW)	Converter (KW)	NPC	Cost of Energy ($)	Operating Cost ($)	Capitol Cost ($/Year)	Renewable Fraction (%)
01	395	100	262		138	250,546	0.0426	719.11	232,568	99.0
02	747		414		155	343,563	0.0584	778.96	324,089	100
03	1218	100			130	691,556	0.118	9100	464,066	97.2
04		100		99,999		1.34 M	0.227	51,458	50,000	0
05		100	100		62.1	2.14 M	0.365	83,004	68,738	0

Table 8.　Economic comparison of proposed five categories of microgrid.

Metric	Category 01	Category 02	Category 03	Category 04	Category 05
Present Worth ($)	1,076,894	983,877	635,884	0	816,408
Annual Worth ($/year)	43,076	39,355	25,435	0	32,656
Return on Investment (%)	22.5	13.9	6.0	0	-117.7
Simple Payback (year)	3.71	5.37	9.56	n/a	n/a

the first optimal design suggested by HOMER exhibits the shortest payback period, the highest return on investment, and the highest present and annual worth among the five proposed designs. These findings emphasize the exceptional suitability of the first proposed microgrid design for the specific site under consideration.

6.3.1. Discussion

In developing nations like Pakistan, power disruptions, high energy costs, conventional energy limitations, and environmental concerns are significant challenges. Addressing these issues through reliable and cost-effective continuous energy supply, primarily from renewable sources, is a global research focus. In our study, we optimized a microgrid for MUST University, Pakistan, using HOMER Pro Software. We analyzed a year's energy usage, then simulated a microgrid with solar, wind, diesel generator, grid, and battery resources, predicting 25-year data. Following 979 simulations, a cost-driven optimal design was identified, meeting the university's energy needs efficiently. Evaluation metrics included Net Present Cost (NPC), Levelized Cost of Energy (COE), Operating Cost (OC), Capital Cost (CC), and Renewable Fraction. Rigorous economic analysis factored capital, fuel, replacement, net present costs, and more, leading to the best microgrid choice. The most effective components were the solar photovoltaic system (SPV), battery bank, diesel generator, and power converter, ensuring continuous backup. SPV dominated energy generation (99.7%), with the diesel generator supplying 0.333% during low sunlight, enhancing reliability. Comparing microgrid scenarios, HOMER's optimal hybrid design demonstrated the shortest payback period, highest return on investment, and superior present and annual worth. This choice excelled among five designs, reinforcing its suitability and cost-efficiency for MUST University's energy needs.

7. Conclusion

In conclusion, this chapter has provided an in-depth exploration of hybrid microgrids and their development in the context of contemporary applications. The chapter began with the definition of hybrid microgrids and highlighted their importance in various sectors. The purpose of the chapter was established, focusing on understanding the applications, energy demand analysis, energy sources, mathematical modelling, and objective parameters for evaluating hybrid microgrids.

The applications of hybrid microgrids were discussed, encompassing remote communities, military installations, industrial and commercial complexes, universities and campuses, disaster resilience, data centers, resorts and tourism facilities, and rural electrification. The analysis of energy demand in these applications was examined, considering factors influencing energy demand and various methods for energy demand analysis.

The chapter further explored the energy sources used in hybrid microgrids, including renewable sources such as solar and wind power, as well as traditional sources like grid connection and diesel generators. The benefits and challenges associated with both renewable and traditional energy sources were discussed, highlighting the importance of balancing the energy mix for optimal performance and sustainability.

Mathematical modelling of selected resources and energy storage systems was explored, emphasizing the role of photovoltaic systems, diesel generators, fuel cells, electrolyzers, hydrogen tanks, and battery storage systems in hybrid microgrid configurations. Objective parameters such as net present cost, levelized cost of electricity, and greenhouse gas emissions were identified as key metrics for evaluating the performance and environmental impact of hybrid microgrids.

Three case studies were presented, examining the cost analysis and greenhouse gas emissions of a cement industry, conducting sensitivity analysis and assessing environmental impacts in the case of Gwadar, and exploring the implementation of hybrid microgrids in a university campus setting.

In conclusion, this chapter has provided valuable insights into the development and evolution of hybrid microgrids, their applications, energy demand analysis, energy sources, mathematical modeling, objective parameters, and case studies. As the field continues to advance, future directions may involve exploring advanced energy storage technologies, optimizing the integration of renewable sources, and developing smart grid functionalities to further enhance the efficiency, reliability, and sustainability of hybrid microgrid systems.

7.1. Future direction

Looking ahead, the future of hybrid microgrids holds great promise for the advancement of sustainable energy systems. As technology continues to evolve, the development of advanced energy storage technologies will play a pivotal role. Research efforts in this area will focus on improving the efficiency and capacity of energy storage solutions, enabling better integration and utilization of intermittent renewable energy sources. Additionally, the optimization of renewable energy integration within hybrid microgrids will be a key area of exploration, aiming to achieve a harmonious balance between different renewable sources to maximize energy generation and minimize reliance on traditional energy sources. The implementation of smart grid functionality will also drive future developments, enabling real-time monitoring, demand response, and automated control systems to enhance the overall performance and efficiency of hybrid microgrids. Furthermore, ensuring the resilience and cybersecurity of hybrid microgrid systems will be a critical consideration, with research and development efforts focusing on designing robust infrastructure and implementing stringent security measures. Ultimately, collaborative efforts among researchers, industry stakeholders, policymakers, and regulatory bodies will be essential in shaping a future where hybrid microgrids play a central role in delivering sustainable and resilient energy solutions.

References

Abdel Gawad, A. E., Alyamani, N. A. and Qaisar, S. M. 2022. Supervisory digital feedback control system for an effective PV management and battery integration, Chapter 9, pp. 165–194. John Wiley & Sons, Ltd. ISBN 9781119816096. URL https://onlinelibrary.wiley.com/doi/abs/10.1002/9781119816096.ch9.

Ali, A., Muqeet, H. A., Khan, T., Hussain, A., Waseem, M. and Niazi, K. A. K. 2023. Iot-enabled campus prosumer microgrid energy management, architecture, storage technologies, and simulation tools: A comprehensive study. Energies, 16(4). ISSN 1996-1073. URL https://www.mdpi.com/1996-1073/16/4/1863.

Ali, M. S., Ali, S. U., Mian Qaisar, S., Waqar, A., Haroon, F. and Alzahrani, A. 2022. Techno-economic analysis of hybrid renewable energy-based electricity supply to gwadar, pakistan. Sustainability, 14(23). ISSN 2071-1050. URL https://www.mdpi.com/2071-1050/14/23/16281.

Alshehri, M. A. H., Guo, Y. and Lei, G. 2023. Renewable-energy-based microgrid design and feasibility analysis for king saud university campus, riyadh. Sustainability, 15(13). ISSN 2071-1050. URL https://www.mdpi.com/2071-1050/15/13/10708.

Ammach, S. and Qaisar, S. M. 2022. A smart energy-efficient support system for pv power plants, Chapter 7, pp. 111–142. John Wiley & Sons, Ltd. ISBN 9781119816096. URL https://onlinelibrary.wiley.com/doi/abs/10.1002/9781119816096.ch7.

Anglani, N., Di Salvo, S. R., Oriti, G. and Julian, A. L. 2023. Steps towards decarbonization of an offshore microgrid: Including renewable, enhancing storage and eliminating need of dump load. Energies, 16(3). ISSN 1996-1073. URL https://www.mdpi.com/1996-1073/16/3/1411.

Awan, M. M. A., Javed, M. Y., Asghar, A. B., Ejsmont, K. and ur Rehman, Z. 2022. Economic integration of renewable and conventional power sources—a case study. Energies, 15(6). ISSN 1996-1073. URL https://www.mdpi.com/1996-1073/15/6/2141.

Ayad, I. A., Elwarraki, E., Ali, S. U., Qaisar, S. M., Waqar, A., Baghdadi, M. and Alzahrani, A. 2023. Optimized nonlinear integral backstepping controller for dc-dc three-level boost converters. IEEE Access, 11: 49794–49805.

Bashawyah, D. A. and Qaisar, S. M. 2021. Machine learning based short-term load forecasting for smart meter energy consumption data in london households. In 2021 IEEE 12th International Conference on Electronics and Information Technologies (ELIT), pp. 99–102.

Basheer, Y., Waqar, A., Qaisar, S. M., Ahmed, T., Ullah, N. and Alotaibi, S. 2022. Analyzing the prospect of hybrid energy in the cement industry of pakistan, using homer pro. Sustainability, 14(19). ISSN 2071-1050. URL https://www.mdpi.com/2071-1050/14/19/12440.

Castellanos, J., Correa-Flórez, C. A., Garcés, A., Ordóñez-Plata, G., Uribe, C. A. and Patino, D. 2023. An energy management system model with power quality constraints for unbalanced multi-microgrids interacting in a local energy market. Applied Energy, 343: 121149. ISSN 0306-2619. URL https://www.sciencedirect.com/science/article/pii/S0306261923005135.

Chisale, S. W. and Mangani, P. 2021. Energy audit and feasibility of solar PV energy system: Case of a commercial building. Journal of Energy. ISSN 2356-735X.

Faheem, M., Butt, R. A., Raza, B., Ashraf, M. W., Ngadi, M. A. and Gungor, V. C. 2019a. A multi-channel distributed routing scheme for smart grid real-time critical event monitoring applications in the perspective of industry 4.0. International Journal of Ad Hoc and Ubiquitous Computing, 32(4): 236–256. URL https://www.inderscienceonline.com/doi/abs/10.1504/IJAHUC.2019.103264.

Faheem, M., Umar, M., Butt, R. A., Raza, B., Ngadi, M. A. and Gungor, V. C. 2019b. Software defined communication framework for smart grid to meet energy demands in smart cities. In 2019 7th International Istanbul Smart Grids and Cities Congress and Fair (ICSG), pp. 51–55.

Faheem, M., Fizza, G., Ashraf, M. W., Butt, R. A., Ngadi, M. A. and Gungor, V. C. 2021. Big data acquired by internet of things-enabled industrial multi-channel wireless sensors networks for active monitoring and control in the smart grid industry 4.0. Data in Brief, 35: 106854. ISSN 2352-3409. URL https://www.sciencedirect.com/science/article/pii/S2352340921001384.

Ghennou, S., Nasser, T. M., Benmoussa, D. and Chellali, B. 2022. Energy optimization of a purely renewable autonomous micro-grid to supply a tourist region. European Journal of Electrical Engineering, 24(1): 33.

Group, T. W. B. 2022. Global solar atlas. URL https://globalsolaratlas.info/map (accessed Jul. 23, 2022).

Hasan, T., Emami, K., Shah, R., Hassan, N., Belokoskov, V. and Ly, M. 2023. Techno-economic assessment of a hydrogen-based islanded microgrid in northeast. Energy Reports, 9: 3380–3396. ISSN 2352-4847. URL https://www.sciencedirect.com/science/article/pii/S2352484723001609.

Icaza-Alvarez, D., Jurado, F., Tostado-Véliz, M. and Arevalo, P. 2022. Design to include a wind turbine and socio-techno-economic analysis of an isolated airplane-type organic building based on a photovoltaic/hydrokinetic/battery. Energy Conversion and Management: X, 14: 100202. ISSN 2590-1745. URL https://www.sciencedirect.com/science/article/pii/S2590174522000253.

Ilahi, T., Izhar, T., Qaisar, S. M., Shami, U. T., Zahid, M., Waqar, A. and Alzahrani, A. 2023. Design and performance analysis of ultra-wide bandgap power devices-based ev fast charger using bi-directional power converters. IEEE Access, 11: 25285–25297.

Kamal, M. M., Ashraf, I. and Fernandez, E. 2023. Optimal sizing of standalone rural microgrid for sustainable electrification with renewable energy resources. Sustainable Cities and Society, 88: 104298. ISSN 2210-6707. URL https://www.sciencedirect.com/science/article/pii/S2210670722006023.

Khan, H., Nizami, I. F., Qaisar, S. M., Waqar, A., Krichen, M. and Almaktoom, A. T. 2022. Analyzing optimal battery sizing in microgrids based on the feature selection and machine learning approaches. Energies, 15(21). ISSN 1996-1073. URL https://www.mdpi.com/1996-1073/15/21/7865.

Li, J., Liu, P. and Li, Z. 2022. Optimal design and techno-economic analysis of a hybrid renewable energy system for off-grid power supply and hydrogen production: A case study of west china. Chemical Engineering Research and Design, 177. ISSN 02638762.

Mian Qaisar, S. 2020. Event-driven coulomb counting for effective online approximation of li-ion battery state of charge. Energies, 13(21): 2020. ISSN 1996-1073. URL https://www.mdpi.com/1996-1073/13/21/5600.

Omotoso, H. O., Al-Shaalan, A. M., Farh, H. M. H. and Al-Shamma'a, A. A. 2022. Technoeconomic evaluation of hybrid energy systems using artificial ecosystem-based optimization with demand side management. Electronics, 11(2). ISSN 2079-9292. URL https://www.mdpi.com/2079-9292/11/2/204.

Pakistan electricity tariff. URL https://nepra.org.pk/consumer%20affairs/Electricity%20Bill.php. [Online].

Prakash, V. J. and Dhal, P. K. 2021. Techno-economic assessment of a standalone hybrid system using various solar tracking systems for kalpeni island, India. Energies, 14(24). ISSN 1996-1073. URL https://www.mdpi.com/1996-1073/14/24/8533.

Qaisar, S. M. 2020. Li-ion battery SoH estimation based on the event-driven sampling of cell voltage. In 2020 2nd International Conference on Computer and Information Sciences (ICCIS), pp. 1–4.

Qaisar, S. M., Bashawyah, D. A., Alsharif, F. and Subasi, A. 2021. A comprehensive review on the application of machine learning techniques for analyzing the smart meter data. Machine Learning for Sustainable Development, 9: 53.

Rao, J., He, Y. and Liu, J. 2023. Standalone versus grid-connected? operation mode and its economic and environmental assessment of railway transport microgrid. Sustainable Cities and Society, 98: 104811. ISSN 2210-6707. URL https://www.sciencedirect.com/science/article/pii/S2210670723004225.

Rehman, S. 2021. Hybrid power systems—sizes, efficiencies, and economics. Energy Exploration and Exploitation, 39. ISSN 20484054.

Reich, D. and Sanchez, S. M. 2023. Sensitivity analysis of hybrid microgrids with application to deployed military units. Naval Research Logistics (NRL), 70(7): 753–769. URL https://onlinelibrary.wiley.com/doi/abs/10.1002/nav.22130.

Shezan, S., Ishraque, M. F., Muyeen, S., Arifuzzaman, S., Paul, L. C., Das, S. K. and Sarker, S. K. 2022. Effective dispatch strategies assortment according to the effect of the operation for an islanded hybrid microgrid. Energy Conversion and Management: X, 14: 100192. ISSN 2590-1745. URL https://www.sciencedirect.com/science/article/pii/S2590174522000150.

Turkdogan, S. 2021. Design and optimization of a solely renewable based hybrid energy system for residential electrical load and fuel cell electric vehicle. Engineering Science and Technology, An International Journal, 24. ISSN 22150986.

Wishart, J., Dong, Z. and Secanell, M. 2006. Optimization of a pem fuel cell system based on empirical data and a generalized electrochemical semi-empirical model. Journal of Power Sources, 161(2): 1041–1055. ISSN 0378-7753. URL https://www.sciencedirect.com/science/article/pii/S0378775306012250.

Zamani Gargari, M., Tarafdar Hagh, M. and Ghassem Zadeh, S. 2023. Preventive scheduling of a multi-energy microgrid with mobile energy storage to enhance the resiliency of the system. Energy, 263: 125597. ISSN 0360-5442. URL https://www.sciencedirect.com/science/article/pii/S0360544222024835.

Zieba Falama, R., Dawoua Kaoutoing, M., Kwefeu Mbakop, F., Dumbrava, V., Makloufi, S., Djongyang, N., Salah, C. B. and Doka, S. Y. 2022. A comparative study based on a techno-environmental-economic analysis of some hybrid grid-connected systems operating under electricity blackouts: A case study in cameroon. Energy Conversion and Management, 251: 114935. ISSN 0196-8904. URL https://www.sciencedirect.com/science/article/pii/S0196890421011110.

Żóładek, M., Figaj, R., Kafetzis, A. and Panopoulos, K. 2023. Energy-economic assessment of self sufficient microgrid based on wind turbine, photovoltaic field, wood gasifier, battery, and hydrogen energy storage. International Journal of Hydrogen Energy. ISSN 0360-3199. URL. https://www.sciencedirect.com/science/article/pii/S0360319923021833.

CHAPTER 9

Formal Methods for Microgrids

Moez Krichen,[a,b,*] *Imed Ben Dhaou,*[c,d,e] *Wilfried Yves Hamilton Adoni*[f,g]
and *Souhir Sghaier*[h]

1. Introduction

The use of DC Microgrid systems has seen significant growth in recent years due
to the need for efficient and sustainable power systems, Shahgholian (2021). The
development of the DC Microgrid has revolutionized power distribution, and it is
supporting the growth of renewable energy sources such as photovoltaic power

[a] Department of Information Technology, Faculty of Computer Science and Information Technology,
    Al-Baha University, Al-Baha 65528, Saudi Arabia.

[b] ReDCAD Laboratory, University of Sfax, Sfax 3038, Tunisia.

[c] Department of Computer Science, Hekma School of Engineering, Computing and Informatics, Dar
    Al-Hekma University, Jeddah P.O. Box 34801, Saudi Arabia.

[d] Department of Computing, University of Turku, 20500 Turku, Finland.

[e] Higher Institute of Computer Sciences and Mathematics, Department of Technology, University of
    Monastir, Monastir 5000, Tunisia.

[f] Helmholtz-Zentrum Dresden-Rossendorf, Center for Advanced Systems Understanding, Untermarkt
    20, 02826 Görlitz, Germany.

[g] Helmholtz-Zentrum Dresden-Rossendorf, Helmholtz Institute Freiberg for Resource Technology,
    Chemnitzer Str. 40, 09599 Freiberg, Germany.

[h] Department of Science and Technology, college of Ranyah, Taif University, P.O. Box 11099, Taif
    21944, Saudi Arabia.

* Corresponding author: moez.krichen@redcad.org, moez.krichen@ieee.org, mkreishan@bu.edu.sa

and wind power. The DC Microgrid systems have some advantages over tradi-tional AC power systems, such as higher efficiency, reduced transmission loss, and the ability to operate in remote areas, Espina et al. (2020). The design of DC Microgrid systems is a challenging task due to the inherent system complexity and variability. Traditional design techniques are becoming inefficient; hence, the need to incorporate formal techniques, Saeed et al. (2021).

As described in the first chapter, there widely reported DC microgrid architec-tures are single-bus, multi-bus, ring-bus, zonal and multi-terminal. Figure 1 il-lustraftes a single-bus microgrid architecture with a microgrid management unit (MMU). Local controllers and power converters communicate with the MMU.

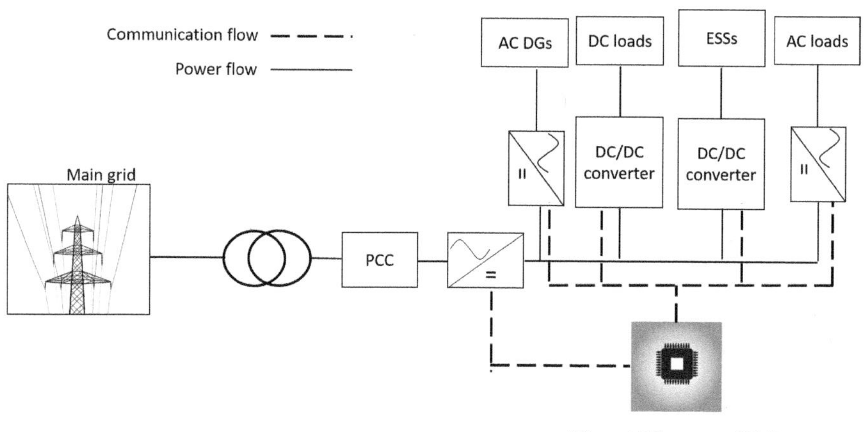

Figure 1. A single bus DC microgrid with a centralized controller.

Despite their potential benefits, microgrids can also be prone to accidents and failures, Mishra et al. (2020); Wu et al. (2021). For example, in 2011, a microgrid at the University of California, San Diego experienced a power outage due to a software error, causing significant disruption to campus operations. Similarly, in 2013, a microgrid in New Jersey failed during Superstorm Sandy, leading to a prolonged power outage and significant economic losses.

The design of a DC microgrid system is a complicated process that must take into account the weather, geographic location, appropriate communication infrastructure, and type of customers. Historically, simulation techniques have been used to test and verify the operation of the microgrid. However, given the number of variables that might effect the reliability and efficiency of the micro-grid, as well as the microgrid's dynamic behavior, researchers have turned to the formal approach to assure the microgrid's appropriate operation, Wu et al. (2022).

Formal methods offer a promising approach to addressing these challenges and minimizing the risk of accidents and failures, Krichen and Tripakis (2006);

Sugumar et al. (2019). Formal methods are mathematical techniques that allow the verification and validation of complex systems, Abdelghany and Tahar (2022). They are particularly useful for designing and managing safety-critical systems, such as microgrids. Formal methods enable engineers to mathematically model and analyze microgrid behaviour, detect errors and inconsistencies, and verify the correctness of system behaviour, Jakaria et al. (2021).

The purpose of this chapter is to explore the use of formal methods for optimizing the performance, safety, and reliability of microgrids. Specifically, the chapter aims to:

- Provide an overview of microgrids and their importance in the context of sustainable energy generation.
- Discuss the benefits of using formal methods for microgrid optimization and the challenges in managing complex energy systems.
- Describe how cloud, fog, and IoT technologies, as well as AI and ML, can be integrated with formal methods to improve microgrid performance.
- Examine the role of formal methods in addressing security concerns in blockchain-based microgrids. Identify the challenges and open issues in this area and recommend future research directions.

The rest of the chapter is structured as follows. Section 2 provides a more detailed overview of microgrids. Section 3 provides an overview of formal methods. Section 4 discusses how cloud technologies can be integrated with formal methods to enhance microgrid performance. Section 5 discusses how fog technologies can be combined with formal methods to enhance microgrid performance. Section 6 describes the benefits of using IoT technologies and formal methods for microgrid optimization, Section 7 discusses how AI and ML can be integrated with formal methods to enhance microgrid safety and reliability. Section 8 discusses the benefits of using formal methods for microgrid security. Section 9 examines the role of formal methods in addressing security concerns in blockchain-based microgrids. Section 10 identifies the challenges and open issues in this area, such as scalability and efficiency of formal methods for large-scale microgrid systems. Section 11 recommends future research directions for the integration of formal methods with microgrid technologies. Finally, Section 12 concludes the chapter.

2. Microgrids

A microgrid is a small-scale power system that can operate independently or parallel with the main grid. Microgrids typically consist of a range of distributed energy resources, such as solar panels, wind turbines, and energy storage systems, and are designed to provide reliable and resilient power to local communities. The concept of microgrids is not new and has been around for decades, but recent advances in technology and the increasing demand for sustainable energy solutions have led to renewed interest in microgrids.

One of the main advantages of microgrids is their ability to operate independently of the main grid. This independence means that microgrids can continue to provide power to local communities even in the event of a power outage or other disruption to the main grid. This reliability and resiliency make microgrids attractive to communities that are located in remote or isolated areas or are vulnerable to natural disasters.

However, microgrids also present certain risks and challenges. One of the main risks associated with microgrids is their vulnerability to cyber-attacks and other security threats. Because microgrids are often connected to the internet or other communication networks, they can potentially be targeted by hackers or other malicious actors. Additionally, microgrids are subject to the same physical risks as any other power system, such as natural disasters or equipment failures.

To protect microgrids from these risks, a range of classical techniques have been developed. These techniques include physical security measures, such as fencing and access control, as well as cybersecurity measures, such as firewalls and intrusion detection systems. Additionally, microgrids can be designed with redundancy and backup systems to ensure that power can be restored quickly in the event of an outage or failure.

However, these classical techniques have their limitations. Physical security measures can be expensive and may not be effective against certain types of threats, such as insider attacks. Cybersecurity measures can also be complex and require specialized knowledge and expertise to implement effectively. Furthermore, these techniques may not be sufficient to protect against emerging threats, such as those posed by artificial intelligence or quantum computing.

To address these limitations, new approaches are needed for protecting microgrids. One promising approach is the use of formal methods, which are a set of mathematical and logical techniques used to design, specify, and verify software and hardware systems. Formal methods provide a rigorous and systematic approach to system design and analysis, which can help to ensure the correctness and reliability of the system.

3. Formal methods techniques

Formal methods are a set of mathematical and logical techniques used to design, specify, and verify software and hardware systems, Krichen et al. (2021). Formally, a method is considered formal if it is based on a mathematical theory, has a well-defined syntax and semantics, and is amenable to rigorous analysis and verification.

As illustrated in Figure 2, the use of formal methods involves the creation of a mathematical model of the system being designed or analyzed. This model is typically expressed in a formal language such as temporal logic or automata theory. The model can then be analyzed using formal methods such as theorem proving or model checking, which involve the use of mathematical algorithms to verify the correctness of the system.

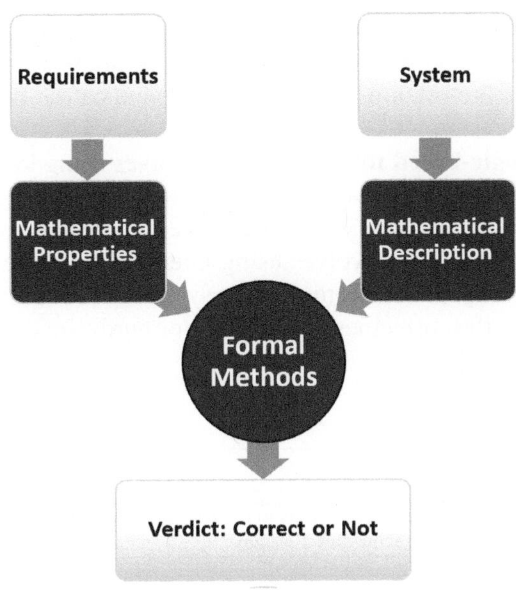

Figure 2. A simplified illustration of how formals methods work.

There are several types of formal methods that are commonly used in the verification and validation of complex systems, including:

- **Theorem proving:** This involves the use of mathematical proofs to demonstrate the correctness of a system. Theorem proving is often used to verify safety-critical systems or systems with complex logic.
- **Model checking:** This involves the use of algorithms to check all possible system behaviors against a formal specification. Model checking is often used to verify finite-state systems or systems with temporal logic.
- **Abstract interpretation:** This involves the use of mathematical abstractions to analyze the behavior of a system. Abstract interpretation is often used to verify programs with complex data structures or non-termination.
- **Runtime verification:** This involves monitoring the behavior of a system during its execution and checking it against a formal specification. Runtime verification is often used to detect and diagnose errors or anomalies in a system's behavior.
- **Model-based testing:** This involves the generation of test cases from a formal model of the system. Model-based testing is often used to ensure the functional correctness of a system.
- **Automata-based methods:** This involves using automata theory to analyze the behavior of a system. Automata-based methods are often used to verify concurrent or distributed systems, aswell as systems with complex control logic.

- **Process-algebras-based methods:** This involves using a formal language to model the behavior of a system as a set of interacting processes. Process algebras are often used to verify communication protocols or other distributed systems.
- **Temporal-logic-based methods:** This involves using logical formulas to describe the temporal behavior of a system. Temporal logic is often used in model checking to verify systems with complex temporal properties.
- **Proof assistants:** This involves using interactive proof assistants to mechanically verify the correctness of a system. Proof assistants are often used to verify the correctness of software or hardware designs.

Table 1. Summary of formal methods.

Type of Formal Method	Description	Common Use Cases
Theorem proving	Use of mathematical proofs to demonstrate the correctness of a system	Safety-critical systems, systems with complex logic
Model checking	Use of algorithms to check all possible system behaviors against a formal specification	Finite-state systems, systems with temporal logic
Abstract interpretation	Use of mathematical abstractions to analyze the behavior of a system	Programs with complex data structures or non-termination
Runtime verification	Monitoring the behavior of a system during its execution and checking it against a formal specification	Detecting and diagnosing errors or anomalies in a system's behavior
Model-based testing	Generation of test cases from a formal model of the system	Ensuring the functional correctness of a system
Automata-based methods	Use of automata theory to analyze the behavior of a system	Concurrent or distributed systems, systems with complex control logic
Process algebras	Use of a formal language to model the behavior of a system as a set of interacting processes	Communication protocols or other distributed systems
Temporal logic	Use of logical formulas to describe the temporal behavior of a system	Model checking for systems with complex temporal properties
Proof assistants	Use of interactive proof assistants to mechanically verify the correctness of a system	Verifying the correctness of software or hardware designs

Each type of formal method has its own strengths and weaknesses, and the appropriate method(s) to use depend on the specific characteristics of the system being verified. By utilizing a combination of different formal methods, engineers can thoroughly analyze and verify the behavior of a system, minimizing the risk of accidents and failures.

To use formal methods to analyze a system, we start by creating a mathematical model of the system. This model is typically expressed in a formal language, which has a well-defined syntax and semantics. The model can then be analyzed using formal methods, such as theorem proving, model checking, or abstract interpretation, to verify the correctness of the system.

The process of analyzing a system using formal methods involves the following steps:

(1) **Formalization:** The first step is to formalize the system by creating a mathematical model that captures its behavior. The model is typically expressed in a formal language, which has a well-defined syntax and semantics.

(2) **Specification:** Once the formal model is created, a specification is developed that describes the desired behavior of the system. The specification is typically expressed in a formal language, which has a well-defined syntax and semantics.

(3) **Verification:** The next step is to verify that the system satisfies the specification. This involves using formal methods, such as theorem proving, model checking, or abstract interpretation, to analyze the formal model and check that it meets the specification.

(4) **Correction:** If the system fails to meet the specification, the next step is to identify the cause of the failure and correct the formal model to address the issue. This may involve revising the model or specification, or modifying the system design.

(5) **Implementation:** Once the formal model and specification are verified, the next step is to implement the system. The implementation is typically done in a programming language, using the formal model and specification as a guide.

(6) **Testing:** Finally, the system is tested to ensure that it behaves correctly. This may involve using techniques such as model-based testing or runtime verification to check that the system meets its specification.

Through this formalization process, formal methods can help to ensure the correctness and reliability of software and hardware systems. By using different types of formal methods such as theorem proving, model checking, abstract interpretation, runtime verification, and model-based testing, we can identify potential design flaws, ensure that the system meets its requirements and specifications, and identify potential security threats and vulnerabilities.

The adoption of formal methods in system design and analysis has been facilitated by the availability of several tools and platforms. Some of the commonly used tools and platforms for formal methods include:

- Coq and Isabelle: These are theorem provers that support formal reasoning and proof development.
- SPIN and UPPAAL: These are model checkers capable of analyzing systems modeled as finite-state machines or timed automata, respectively.
- Java PathExplorer and T-VEC: These are tools for runtime verification and model-based testing of Java programs and Simulink models, respectively.

- Alloy and TLA+: These are languages and tools for modeling and analyzing software systems using relational logic and temporal logic, respectively.
- Frama-C and VeriFast: These are platforms and tools for analyzing and verifying C programs using a range of formal methods.
- CBMC: This is a bounded model checker for verifying C and C++ programs using a combination of model checking and theorem proving.
- NuSMV: This is a symbolic model checker for verifying systems modeled as finite-state machines or transition systems.
- Z3: This is a theorem prover and SMT solver that supports a range of logics and theories, including arithmetic, bit-vectors, and arrays.
- Why3: This is a platform and toolset for deductive program verification that supports a range of programming languages and verification tools.

Table 2. Summary of formal method tools and platforms.

Tool/Platform	Description	Common Use Cases
Coq and Isabelle	Theorem provers that support formal reasoning and proof development	Verifying safety-critical systems or systems with complex logic
SPIN and UPPAAL	Model checkers capable of analyzing systems modeled as finite-state machines or timed automata, respectively	Verifying finite-state systems or systems with temporal logic
Java PathExplorer and T-VEC	Tools for runtime verification and model-based testing of Java programs and Simulink models, respectively	Ensuring the functional correctness of Java programs or Simulink models
Alloy and TLA+	Languages and tools for modeling and analyzing software systems using relational logic and temporal logic, respectively	Modeling and analyzing software systems with complex data structures or temporal properties
Frama-C and VeriFast	Platforms and tools for analyzing and verifying C programs using a range of formal methods	Verifying the correctness of C programs
CBMC	Bounded model checker for verifying C and C++ programs using a combination of model checking and theorem proving	Verifying the correctness of C and C++ programs
NuSMV	Symbolic model checker for verifying systems modeled as finite-state machines or transition systems	Verifying finite-state systems or systems with complex control logic
Z3	Theorem prover and SMT solver that supports a range of logics and theories, including arithmetic, bit-vectors, and arrays	Verifying the correctness of software or hardware designs using a range of logics and theories
Why3	Platform and toolset for deductive program verification that supports a range of programming languages and verification tools	Verifying the correctness of programs written in a variety of programming languages

These tools and platforms provide engineers with a range of options for applying formal methods to verify and validate complex systems. By using these tools in conjunction with formal methods, engineers can thoroughly analyze and verify the behavior of a system, minimizing the risk of accidents and failures.

Formal methods offer several advantages for system design and analysis, including:

- Rigorous and systematic approach: Formal methods provide a rigorous and systematic approach to system design and analysis, which can help to ensure the correctness and reliability of the system.
- Early error detection: Formal methods can help to identify potential errors or defects in the system's behavior early in the design process before the system is implemented.
- Improved system performance: Formal methods can help to optimize system performance by identifying potential bottlenecks or areas for improvement in the system's design.
- Enhanced security: Formal methods can help to identify potential security threats and vulnerabilities in the system's design and develop effective defense strategies.

As the complexity and criticality of software and hardware systems continue to increase, the use of formal methods will become increasingly important for ensuring their safety, reliability, and security.

4. Optimizing microgrids with cloud technologies: Ensuring safety and reliability through formal methods

The management and optimization of microgrids are becoming more dependent on cloud technologies. Microgrids produce large volumes of data, including records of energy generation and consumption, which can be stored, processed, and analyzed using cloud computing and storage resources, Dabbaghjamanesh et al. (2020); Tajalli et al. (2019). As a result, microgrid operators will be able to improve the efficiency of their system and save money on energy costs, Wang et al. (2020).

However, questions concerning the security and dependability of cloud technologies are also raised by their implementation in microgrids. For data to be safely stored and processed in the cloud, the underlying infrastructure must be impenetrable to hackers. In such cases, formal approaches are necessary. When it comes to ensuring the security of software systems, especially cloud infrastructure, nothing beats the mathematical rigor of formal approaches. They can be used to guarantee the integrity and security of the microgrid's cloud infrastructure.

It is possible to apply formal methods to verify the correct operation of cloud infrastructure in every scenario and to identify any flaws or security holes, Chen et al. (2019). Furthermore, formal approaches can be used to produce detailed requirements for cloud systems that can be used to direct the development process and guarantee that the final result fulfils all expectations. Using formal methodologies, designers may ensure the reliability of the microgrid's cloud infrastructure. This can help avoid crises caused by things like power outages or broken machinery, Muniasamy et al. (2019).

In addition, the cloud infrastructure utilized by the microgrid system can be made secure with the help of formal approaches. The security of the cloud infrastructure may be checked for flaws that hackers could exploit with the help of formal methodologics. This can aid in preventing hacking and other forms of cybercrime.

In addition, formal methods offer a means of analyzing and bettering the efficacy of the cloud infrastructure, both of which contribute to optimizing the microgrid system's performance. Developers of the microgrid system can fix performance issues and find places for growth in the underlying cloud infrastructure by employing formal methodologies. Microgrid operators may experience greater energy savings and lower operating expenses as a result of this.

When formal approaches are combined with cloud computing, the resulting microgrid system is one that is more secure, reliable, and cost-effective. To minimize the risk of accidents and cyberattacks, and maximize the efficiency of the microgrid system, formal approaches can be used to guarantee the security and dependability of cloud infrastructure.

5. Enhancing microgrid performance with fog technologies and formal methods

Fog computing is a type of distributed computing that brings cloud computing to the edge of the network, where data-generating and -using devices and resources are located. The general scheme of a Fog Computing Architecture is illustrated in Figure 3. With the help of fog technologies, microgrids can collect and analyze data in real-time, which cuts down on latency and boosts performance, Barros et al. (2019). To better manage energy consumption and distribution in real-time, for instance, fog devices can gather and analyse data from Internet of Things (IoT) devices and sensors deployed within the microgrid, Keskin and İnce (2022).

However, questions about the security and dependability of fog technology in microgrids have been raised. The data storage and processing fog infrastructure must be secure and free from vulnerabilities. In such cases, formal approaches are necessary. The validity of software systems, including fog infrastructure, can be verified with the help of formal methods. They can be used to make sure the microgrid's fog infrastructure is secure and dependable.

Proof of correct operation in all scenarios and the identification of flaws and vulnerabilities in fog infrastructure can be accomplished with the use of formal approaches. In addition, formal approaches can be utilized to produce detailed

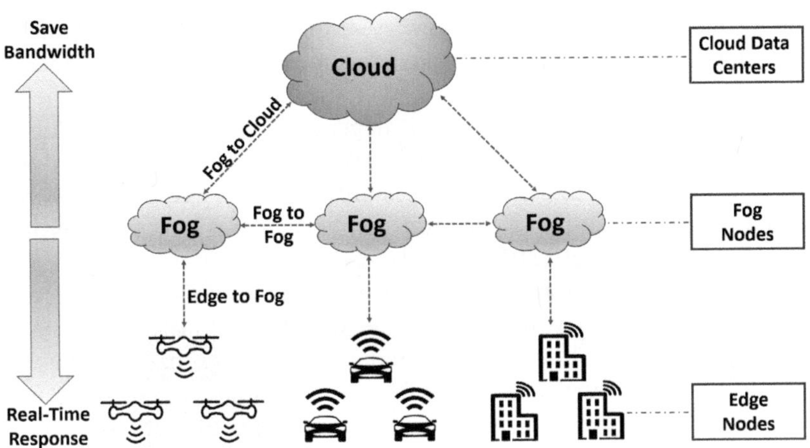

Figure 3. Fog computing general scheme.

requirements for fog systems that can be used to direct and verify the development process. The developers of the microgrid system can rest easy knowing their fog infrastructure is error-free thanks to formal methods of verification, Benzadri et al. (2021). This can help avoid crises caused by things like power outages or broken machinery, Marir et al. (2022).

In addition, the microgrid's fog infrastructure can be made secure with the help of formal approaches. The absence of flaws that could be exploited by cybercriminals in the fog infrastructure can be verified with the help of formal approaches. This can aid in preventing [hacki]ng and other forms of cybercrime.

In addition, formal approaches offer a means of analyzing and bettering the efficacy of the fog infrastructure, both of which contribute to optimizing the microgrid system's performance. Developers of the microgrid system's fog infrastructure can find potential performance concerns and areas for improvement by employing formal methodologies. Microgrid operators may experience greater energy savings and lower operating expenses as a result of this.

An improved microgrid in terms of security, dependability, and efficiency can be achieved through the employment of formal approaches in tandem with fog technologies. To minimize the risk of accidents and cyberattacks, and maximize the efficiency of the microgrid, formal approaches can be used to guarantee the security and dependability of fog infrastructure.

6. Ensuring safety and reliability of microgrids with IoT and formal methods

Microgrids increasingly utilize Internet of Things (IoT) technologies, Gonzalez et al. (2022); Saad et al. (2020). Internet of Things devices can track and manage microgrid operations like energy generation, distribution, and storage. Sensors can monitor the output of renewable energy sources like solar panels and wind turbines, allowing the microgrid to make adjustments that maximize efficiency, Lei et al. (2020); Silva et al. (2023).

Concerns concerning the security and dependability of IoT technology are also raised by their deployment in microgrids. Safety issues or power outages may occur if the Internet of Things devices utilized in the microgrid system are malfunctioning or susceptible to cyber threats. In such cases, formal approaches are necessary, Oliveira et al. (2021).

Proof of correct operation in all scenarios, as well as the identification of mistakes and vulnerabilities, are all possible with the help of formal methods applied to the Internet of Things. Rigid specifications for Internet of Things (IoT) systems can be created with the use of formal methodologies, which can then be used to direct the development process and guarantee that the final product fulfills all criteria. Using formal approaches, developers may ensure the integrity and accuracy of the microgrid's IoT devices. This can help avoid crises caused by things like power outages or broken machinery, Hofer-Schmitz and Stojanović (2019).

In addition, formal approaches can aid in optimizing the microgrid system's performance by offering a means to examine and enhance the efficiency of the IoT devices. Using formal methods, designers can probe the microgrid IoT devices for performance flaws and optimization opportunities. Microgrid operators may experience greater energy savings and lower operating expenses as a result of this, Webster et al. (2020).

IoT devices utilized in the microgrid infrastructure can have their security guaranteed by employing formal approaches. It is possible to utilize formal procedures to ensure that the IoT devices are secure from any flaws that could be leveraged by hackers. This can aid in preventing hacking and other forms of cybercrime.

When formal approaches are used with Internet of Things technologies, the resulting microgrid can be more secure, dependable, and efficient. Formal approaches can help to optimize the performance of the microgrid system by ensuring the safety and dependability of IoT devices, hence preventing potential safety concerns and cyber threats.

7. AI and ML for microgrids: Ensuring safety and reliability with formal methods

AI (Artificial Intelligence) and ML (Machine Learning) are increasingly being used for microgrids, Etingov et al. (2022); Zhou et al. (2023). The difference between the two concepts is illustrated in Figure 4. By using AI and ML, microgrids can optimize their energy usage and reduce their carbon footprint, Puerta et al. (2023). For example, AI can be used to monitor and predict energy demand, as well as to control and manage energy storage systems. This means that micro-

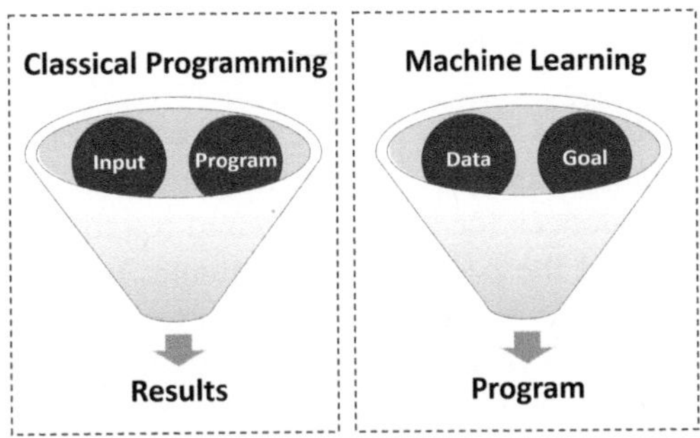

Figure 4. The difference between classical programming and machine learning.

grids can adjust their energy output based on real-time demand and supply data, reducing energy waste and increasing efficiency, Mohammadi et al. (2022). ML can be used for energy forecasting, anomaly detection, and predictive maintenance. This means that microgrids can predict future energy demand and detect any abnormalities or faults in the system, reducing the risk of power outages or equipment failures.

However, the use of AI and ML in microgrids also raises concerns about their safety and reliability. If the algorithms used in the microgrid system are not correct or free from errors, it could potentially lead to safety hazards or power outages.

Formal methods can be used to prove that AI and ML algorithms will work correctly in all situations, and to detect any potential errors or vulnerabilities, Krichen et al. (2022b). In addition, formal methods can be used to generate rigorous specifications for AI and ML systems, which can be used to guide the development process and ensure that the final product meets all requirements, Adjed et al. (2022). By using formal methods, developers can verify that the AI and ML algorithms used in the microgrid system are correct and free from errors, Al-Nusair (2020). This can help to prevent potential safety hazards, such as power outages or equipment failures, Larsen et al. (2022).

Moreover, formal methods can help to optimize the performance of the microgrid system by providing a way to analyze and improve the efficiency of the AI and ML algorithms, Gossen et al. (2020). By using formal methods, developers can identify potential performance issues or areas for improvement in the microgrid system, Huang et al. (2022). This can result in greater energy savings and reduced costs for microgrid operators. Overall, the use of formal methods in conjunction with AI and ML can lead to a safer, more reliable, and more efficient microgrid system.

In conclusion, AI and ML have great potential for optimizing the energy usage of microgrids, but their use also requires careful consideration of safety and reliability concerns. The use of formal methods can help to ensure that the AI and ML algorithms used in the microgrid system are free from errors and meet all requirements, resulting in a safer and more efficient microgrid system.

8. Formal methods for security aspects

Microgrids are vulnerable to a variety of security threats, including cyber attacks, physical attacks, and natural disasters, Nejabatkhah et al. (2020). These threats can have serious consequences for the stability and reliability of the microgrid, as well as for the safety of the people and equipment involved. The use of formal methods can play an important role in ensuring the security of microgrids by providing a rigorous and systematic approach to analyzing and verifying the security properties of the system, Krichen et al. (2018, 2020).

Formal methods can be used to model and analyze the security properties of a microgrid, including the identification of potential security vulnerabilities and the development of countermeasures to address these vulnerabilities. Formal methods can also be used to verify the correctness of security protocols and algorithms used in the microgrid, as well as to ensure that the system is resilient to attacks and can recover quickly in the event of a security breach.

Adapting formal methods to the specific security needs of microgrids requires an understanding of the unique characteristics of these systems. Microgrids are typically composed of many interconnected components, including power sources, loads, and monitoring and control systems. These components may be owned and operated by different entities, which can make it challenging to ensure consistent security practices across the system. Additionally, microgrids may be subject to rapid changes inpower supply and demand, which can make it challenging to model and analyze the system accurately.

To address these challenges, stakeholders in the field of microgrid security can adapt existing formal methods to the specific needs of microgrids. This may involve developing new modeling techniques and tools that can capture the unique characteristics of microgrids, as well as developing new verification and validation methods that can be applied to these models.

In summary, the use of formal methods can play an important role in ensuring the security of microgrids. Adapting formal methods to the specific needs of microgrids requires an understanding of the unique characteristics of these systems and the development of new techniques and tools to address these characteristics. Additionally, addressing the human factors involved in the use of formal methods for security is crucial for ensuring that the security needs of microgrids are understood and addressed in a coordinated and consistent manner. By addressing these issues, stakeholders can ensure that microgrids are secure, reliable, and resilient in the face of increasing security threats, helping to ensure the safety and stability of the power grid as a whole.

9. Blockchain-based microgrids: The role of formal methods

Blockchain technology and smart contracts have emerged as promising tools for the design and operation of microgrids, Krichen et al. (2022a); Li et al. (2019).

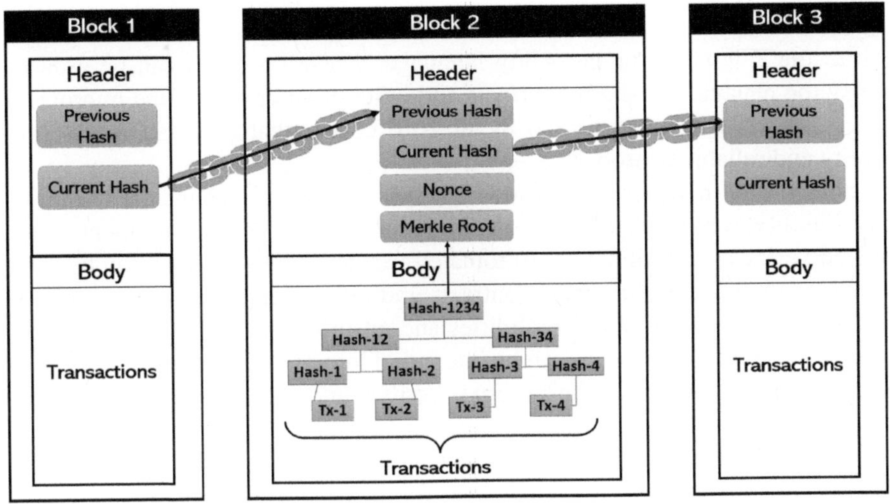

Figure 5. Blockchain general architecture.

Blockchain technology provides a decentralized and secure platform for recording transactions and managing distributed systems, while smart contracts enable the automation of transactions and the execution of complex business logic, Dinesha and Balachandra (2022); Tsao and Thanh (2021). The general architecture of blockchain is shown in Figure 5.

The use of blockchain technology and smart contracts in microgrids presents a number of security and reliability challenges, however. These challenges arise from the distributed and open nature of blockchain systems, as well as the complexity of the smart contracts that govern their behavior. Formal methods can be used to address these challenges and help to ensure the security and reliability of blockchain-based microgrids.

Formal methods can be used to model and analyze the behavior of blockchain systems and smart contracts, and to verify that they satisfy a set of security and reliability properties, Murray and Anisi (2019). Formal methods can also be used to identify potential vulnerabilities in the system and to develop countermeasures to mitigate these vulnerabilities, Brunese et al. (2019).

The use of formal methods in the context of blockchain-based microgrids is still a relatively new area of research, but holds great promisefor ensuring the security and reliability of these systems, Brunese et al. (2019). Formal methods can be used to verify the correctness of smart contracts that govern the behavior of microgrid components, ensuring that they operate as expected and that they do not introduce vulnerabilities into the system. Formal methods can also be used to analyze the behavior of the microgrid as a whole, identifying potential vulnerabilities and ensuring that the system is resilient to attacks.

One way in which formal methods can be used to enhance the security and reliability of blockchain-based microgrids is through the development of formal contracts. Formal contracts are contracts that are expressed in a formal language, such as the Z notation or the Alloy language. These contracts can be used to specify the behavior of smart contracts and other components of the microgrid, and can be verified using formal methods to ensure that they satisfy a set of security and reliability properties.

Another way in which formal methods can be used to enhance the security and reliability of blockchain-based microgrids is through the development of automated verification techniques. Automated verification techniques can be used to analyze the behavior of smart contracts and other components of the microgrid, identifying potential vulnerabilities and ensuring that the system operates as expected. These techniques can be integrated into the development process of the microgrid, ensuring that the system is verified and validated at every stage of development.

In summary, the integration of blockchain technology and smart contracts presents both opportunities and challenges for the design and operation of microgrids. The use of formal methods can help to ensure the security and reliabilityof these systems, by providing a rigorous and systematic approach to the analysis and verification of their behavior. Formal methods can be used to model and analyze the behavior of blockchain systems and smart contracts, and to verify that they satisfy a set of security and reliability properties. The development of formal contracts and automated verification techniques can further enhance the security and reliability of blockchain-based microgrids.

10. Challenges and open issues

The use of formal methods in microgrid design and analysis is an emerging area of research that holds great promise for improving the safety, reliability, and efficiency of these systems. However, there are several challenges and open issues that must be addressed to fully realize the potential of formal methods in microgrid design and analysis. These challenges and open issues include:

- **Model complexity:** Microgrids can be composed of various types of power sources and can be difficult to model accurately. The interaction between these different sources can be complex and difficult to capture, which makes it challenging to apply formal methods to analyze the behavior of the system.
- **Lack of standardized models and specifications:** There is no widely accepted standard for modeling and specifying microgrids. This makes it difficult to compare different microgrid designs and to apply formal methods consistently across different systems.
- **Dynamic nature of microgrids:** Microgrids can be subject to rapid changes in power supply and demand, as well as fluctuations in weather and environmental conditions. These changes can make it difficult to develop accurate models and specifications for the system, and can also

make it challenging to apply formal methods to analyze the behavior of the system in real-time.

- **Computational complexity:** Formal methods often require significant computational resources to analyze even small systems, which can be a significant barrier to their adoption in microgrid design and analysis. This is particularly true for model checking, which can require large amounts of memory and processing power to analyze complex systems.
- **Practical considerations:** The cost of developing and verifying formal models for microgrids can be significant, particularly for small-scale systems. This can limit the adoption of formal methods in microgrid design and analysis, particularly for systems with limited budgets or resources.
- **Verification and validation:** Verification and validation of formal models can be challenging, particularly for complex and dynamic systems such as microgrids. There is a need for methods and tools to ensure that formal models accurately capture the behavior of the system and can be verified and validated in a timely and efficient manner.
- **Human factors:** The use of formal methods in microgrid design and analysis requires collaboration between technical experts and stakeholders such as system operators and policymakers. There is a need for methods and tools that can facilitate communication and collaboration between these different groups to ensure that the formal models accurately capture the needs and requirements of the system.
- **Real-world implementation:** The implementation of formal models in real-world microgrid systems can be challenging, particularly in systems that have already been designed and deployed. There is a need for methods and tools that can facilitate the integration of formal models into existing systems and workflows, as well as methods to ensure that the models continue to accurately reflect the behavior of the system over time.

By addressing these challenges and open issues, formal methods have the potential to play an important role in the design, operation, and maintenance of microgrids, helping to ensure their safety, reliability, and efficiency in the face of increasing demand for sustainable and resilient power systems.

11. Future directions and recommendations

As the field of microgrid design and analysis continues to evolve, the use of formal methods is poised to play an increasingly important role in ensuring the safety, reliability, and efficiency of these systems. However, to fully realize the potential of formal methods in microgrid design and analysis, there are several future directions and recommendations that stakeholders in this field should consider:

- **Develop standardized models and specifications:** The development of standardized models and specifications for microgrids is crucial for the adoption and application of formal methods in this area. Stakeholders in this field should work together to develop and promote widely accepted standards for microgrid modeling and specification to facilitate the use of formal methods across different systems.
- **Address computational complexity:** The computational complexity of formal methods can be a significant barrier to their adoption in microgrid design and analysis. Stakeholders should focus on developing and promoting methods and tools that can reduce the computational burden of formal methods, such as model reduction techniques and efficient model checking algorithms.
- **Integrate formal methods into existing workflows:** To be effective, formal methods must be integrated into existing microgrid design and analysis workflows. Stakeholders should focus on developing and promoting methods and tools that can facilitate the integration of formal methods into existing workflows and systems, as well as methods to ensure that the models continue to accurately reflect the behavior of the system over time.
- **Address human factors:** The successful application of formal methods in microgrid design and analysis requires collaboration between technical experts and stakeholders such as system operators and policymakers. Stakeholders should focus on developing and promoting methods and tools that can facilitate communication and collaboration between these different groups to ensure that the formal models accurately capture the needs and requirements of the system.
- **Validate and verify formal models:** The verification and validation of formal models is crucial for ensuring that they accurately capture the behavior of the system. Stakeholders should focus on developing and promoting methods and tools that can efficiently and effectively verify and validate formal models, as well as methods for ensuring that the models remain accurate over time.
- **Develop practical solutions:** The cost of developing and verifying formal models for microgrids can be significant, particularly for small-scale systems. Stakeholders should focus on developing and promoting practical solutions that can be adopted by systems with limited budgets or resources, such as simplified modeling techniques and efficient verification and validation methods.
- **Promote collaboration and knowledge sharing:** The successful adoption of formal methods in microgrid design and analysis requires collaboration and knowledge sharing across disciplines and sectors. Stakeholders should focus on promoting collaboration and knowledge sharing through conferences, workshops, and other events, as well as through the development of onlinecommunities and resources for sharing best practices and lessons learned.

- **Address ethical and social implications:** The use of formal methods in microgrid design and analysis raises important ethical and social implications, such as issues related to privacy, fairness, and accountability. Stakeholders should be aware of these implications and work to address them in the development and application of formal methods in microgrid design and analysis.

In summary, the use of formal methods in microgrid design and analysis holds great promise for improving the safety, reliability, and efficiency of these systems. To fully realize this potential, stakeholders in this field should focus on developing standardized models and specifications, addressing computational complexity, integrating formal methods into existing workflows, addressing human factors, validating and verifying formal models, developing practical solutions, promoting collaboration and knowledge sharing, and addressing ethical and social implications. By addressing these future directions and recommendations, stakeholders can ensure that formal methods play an increasingly important role in the design, operation, and maintenance of microgrids, helping to ensure their safety, reliability, and efficiency in the face of increasing demand for sustainable and resilient power systems.

12. Conclusion

In this chapter, we have explored the use of formal methods for optimizing the performance, safety, and reliability of microgrids. Microgrids are becoming increasingly popular as a means of providing reliable and sustainable energy, but managing their complexity can be challenging. Formal methods offer a promising approach to addressing this challenge by enabling the verification and validation of complex systems.

We have discussed various technologies that can be integrated with formal methods to enhance the performance and safety of microgrids. These include cloud, fog, and IoT technologies, as well as AI and ML. By integrating these technologies with formal methods, microgrid operators can optimize energy generation and consumption, prevent power outages, and ensure the safe and reliable operation of their systems.

We have also discussed the role of formal methods in addressing security concerns in blockchain-based microgrids. Formal methods can help ensure the integrity and confidentiality of transactions in these systems, as well as prevent attacks on the microgrid infrastructure.

Despite the promise of formal methods, there are still several challenges and open issues that must be addressed. For example, there is a need for more research on the scalability and efficiency of formal methods for large-scale microgrid systems. Additionally, there is a need for more standardized formal methods and tools that can be used by microgrid operators and engineers.

In conclusion, the use of formal methods for microgrid optimization is a promising approach to addressing the challenges of managing complex energy systems. By integrating formal methods with cloud, fog, and IoT technologies, as well as AI and ML, microgrid operators can ensure the safe and reliable operation of their systems while optimizing energy generation and consumption. Additionally, formal methods can help address security concerns in blockchain-based microgrids. However, further research is needed to address the challenges and open issues in this area and to develop standardized formal methods and tools. Overall, the use of formal methods represents an important step forward in the development of sustainable and reliable microgrid systems.

References

Abdelghany, M. and Tahar, S. 2022. Reliability analysis of smart grids using formal methods. Handbook of Smart Energy Systems. Springer, 1–17.

Adjed, F., Mziou-Sallami, M., Pelliccia, F., Rezzoug, M., Schott, L., Bohn, C. and Jaafra, Y. 2022. Coupling algebraic topology theory, formal methods and safety requirements toward a new coverage metric for artificial intelligence models. Neural Computing and Applications, 34(19): 17129–17144.

Al-Nusair, R. G. 2020. How can factors underlying human preferences lead to methods of formal characterizations towards developing safe artificial general intelligence?

Barros, E. B. C., Filho, D. M. L., Batista, B. G., Kuehne, B. T. and Peixoto, M. L. M. 2019. Fog computing model to orchestrate the consumption and production of energy in microgrids. Sensors, 19(11): 2642.

Benzadri, Z., Bouheroum, A. and Belala, F. 2021. A formal framework for secure fog architectures: Application to guarantee reliability and availability. International Journal of Organizational and Collective Intelligence (IJOCI), 11(2): 51–74.

Brunese, L., Mercaldo, F., Reginelli, A. and Santone, A. 2019. A blockchain based proposal for protecting healthcare systems through formal methods. Procedia Computer Science, 159: 1787–1794.

Chen, S., Fan, Z., Shen, H. and Feng, L. 2019. Performance modeling and verification of load balancing in cloud systems using formal methods. In 2019 IEEE 16th International Conference on Mobile ad hoc and Sensor Systems Workshops (MASSW), pp. 146–151.

Dabbaghjamanesh, M., Kavousi-Fard, A. and Dong, Z. Y. 2020. A novel distributed cloud-fog based framework for energy management of networked microgrids. IEEE Transactions on Power Systems, 35(4): 2847–2862.

Dinesha, D. L. and Balachandra, P. 2022. Conceptualization of blockchain enabled interconnected smart microgrids. Renewable and Sustainable Energy Reviews, 168: 112848.

Espina, E., Llanos, J., Burgos-Mellado, C., Cardenas-Dobson, R., Martinez-Gomez, M. and Saez, D. 2020. Distributed control strategies for microgrids: An overview. IEEE Access, 8: 193412–193448.

Etingov, D. A., Zhang, P., Tang, Z. and Zhou, Y. 2022. Ai-enabled traveling wave protection for microgrids. Electric Power Systems Research, 210: 108078.

Gonzalez, I., Calderón, A. J. and Folgado, F. J. 2022. Iot real time system for monitoring lithium-ion battery long-term operation in microgrids. Journal of Energy Storage, 51: 104596.

Gossen, F., Margaria, T. and Steffen, B. 2020. Towards explainability in machine learning: The formal methods way. IT Professional, 22(4): 8–12.

Hofer-Schmitz, K. and Stojanović, B. 2019. Towards formal methods of iot application layer protocols. In 2019 12th CMI Conference on Cybersecurity and Privacy (CMI), pp. 1–6.

Huang, X., Ruan, W., Tang, Q. and Zhao, X. 2022. Bridging formal methods and machine learning with global optimisation. In Formal Methods and Software Engineering: 23rd International Conference on Formal Engineering Methods, icfem 2022, Madrid, Spain, October 24–27, 2022, Proceedings, pp. 1–19.

Jakaria, A., Rahman, M. A. and Gokhale, A. 2021. Resiliency-aware deployment of sdn in smart grid scada: A formal synthesis model. IEEE Transactions on Network and Service Management, 18(2): 1430–1444.

Keskin, T. and İnce, G. 2022. An ensemble learning approach for energy demand forecasting in microgrids using fog computing. In Intelligent and Fuzzy Techniques for Emerging Conditions and Digital Transformation: Proceedings of the Infus 2021 Conference, held August 24–26, 2021, 2: 170–178.

Krichen, M. and Tripakis, S. 2006. Interesting properties of the real-time conformance relation tioco. In Theoretical Aspects of Computing-ictac 2006: Third International Colloquium, Tunis, Tunisia, November 20–24, 2006. Proceedings 3, 317–331.

Krichen, M., Ammi, M., Mihoub, A. and Almutiq, M. 2022. Blockchain for modern applications: A survey. Sensors, 22(14): 5274.

Krichen, M., Cheikhrouhou, O., Lahami, M., Alroobaea, R. and Jmal Maâlej, A. 2018. Towards a model-based testing framework for the security of internet of things for smart city applications. In Smart Societies, Infrastructure, Technologies and Applications: First International Conference, Scita 2017, Jeddah, Saudi Arabia, November 27–29, 2017, Proceedings 1: 360–365.

Krichen, M., Lahami, M., Cheikhrouhou, O., Alroobaea, R., and Maâlej, A. J. 2020. Security testing of internet of things for smart city applications: A formal approach. In Smart Infrastructure and Applications, pp. 629–653. Springer, Cham.

Krichen, M., Mechti, S., Alroobaea, R., Said, E., Singh, P., Khalaf, O. I. and Masud, M. 2021. A formal testing model for operating room control system using internet of things. Computers, Materials & Continua, 66(3): 2997–3011.

Krichen, M., Mihoub, A., Alzahrani, M. Y., Adoni, W. Y. H. and Nahhal, T. 2022. Are formal methods applicable to machine learning and artificial intelligence? In

2022 2nd International Conference of Smart Systems and Emerging Technologies (smarttech), pp. 48–53.

Larsen, K., Legay, A., Nolte, G., Schlüter, M., Stoelinga, M. and Steffen, B. 2022. Formal methods meet machine learning (f3ml). In Leveraging Applications of Formal Methods, Verification and Validation, Adaptation and Learning: 11th International Symposium, Isola 2022, Rhodes, Greece, October 22–30, 2022, Proceedings, Part III, pp. 393–405.

Lei, L., Tan, Y., Dahlenburg, G., Xiang, W. and Zheng, K. 2020. Dynamic energy dispatch based on deep reinforcement learning in iot-driven smart isolated microgrids. IEEE Internet of Things Journal, 8(10): 7938–7953.

Li, J., Zhou, Z., Wu, J., Li, J., Mumtaz, S., Lin, X., ... Alotaibi, S. 2019. Decentralized on-demand energy supply for blockchain in internet of things: A microgrids approach. IEEE Transactions on Computational Social Systems, 6(6): 1395–1406.

Marir, S., Belala, F. and Hameurlain, N. 2022. A strategy-based formal approach for fog systems analysis. Future Internet, 14(2): 52.

Mishra, S., Anderson, K., Miller, B., Boyer, K. and Warren, A. 2020. Microgrid resilience: A holistic approach for assessing threats, identifying vulnerabilities, and designing corresponding mitigation strategies. Applied Energy, 264: 114726.

Mohammadi, H., Mohammadi, M., Kavousifard, A. and Dabbaghjamanesh, M. 2022. Ai-based optimal scheduling of renewable ac microgrids with bidirectional lstm-based wind power forecasting. arXiv preprint arXiv:2208.04156.

Muniasamy, K., Srinivasan, S., Vain, J. and Sethumadhavan, M. 2019. Formal methods based security for cloud-based manufacturing cyber physical system. IFAC-PapersOnLine, 52(13): 1198–1203.

Murray, Y. and Anisi, D. A. 2019. Survey of formal verification methods for smart contracts on blockchain. In 2019 10th IFIP International Conference on New Technologies, Mobility and Security (NTMS), pp. 1–6.

Nejabatkhah, F., Li, Y. W., Liang, H. and Reza Ahrabi, R. 2020. Cyber-security of smart microgrids: A survey. Energies, 14(1): 27.

Oliveira, L. P., da Silva, A. W. N., de Azevedo, L. P. and da Silva, M. V. L. 2021. Formal methods to analyze energy efficiency and security for IoT: A systematic review. In Advanced Information Networking and Applications: Proceedings of the 35th International Conference on Advanced Information Networking and Applications (aina-2021), volume 3 (pp. 270–279).

Puerta, A., Hoyos, S. H., Bonet, I. and Caraffini, F. 2023. An AI-based support system for microgrids energy management. In Applications of Evolutionary Computation: 26th European Conference, Evoapplications 2023, held as part of Evostar 2023, Brno, Czech Republic, April 12–14, 2023, Proceedings, pp. 507–518.

Saad, A., Faddel, S., Youssef, T. and Mohammed, O. A. 2020. On the implementation of IoT-based digital twin for networked microgrids resiliency against cyber attacks. IEEE Transactions on Smart Grid, 11(6): 5138–5150.

Saeed, M. H., Fangzong, W., Kalwar, B. A. and Iqbal, S. 2021. A review on microgrids' challenges & perspectives. IEEE Access, 9: 166502–166517.

Shahgholian, G. 2021. A brief review on microgrids: Operation, applications, modeling, and control. International Transactions on Electrical Energy Systems, 31(6): e12885.

Silva, J. A. A., López, J. C., Guzman, C. P., Arias, N. B., Rider, M. J. and da Silva, L. C. 2023. An IoT-based energy management system for AC microgrids with grid and security constraints. Applied Energy, 337: 120904.

Sugumar, G., Selvamuthukumaran, R., Novak, M. and Dragicevic, T. 2019. Supervisory energy-management systems for microgrids: Modeling and formal verification. IEEE Industrial Electronics Magazine, 13(1): 26–37.

Tajalli, S. Z., Tajalli, S. A. M., Kavousi-Fard, A., Niknam, T., Dabbaghjamanesh, M. and Mehraeen, S. 2019. A secure distributed cloud-fog based framework for economic operation of microgrids. In 2019 IEEE Texas Power and Energy Conference (TPEC), pp. 1–6.

Tsao, Y. -C. and Thanh, V. -V. 2021. Toward sustainable microgrids with blockchain technology-based peer-to-peer energy trading mechanism: A fuzzy meta-heuristic approach. Renewable and Sustainable Energy Reviews, 136: 110452.

Wang, S., Wang, X. and Wu, W. 2020. Cloud computing and local chip-based dynamic economic dispatch for microgrids. IEEE Transactions on Smart Grid, 11(5): 3774–3784.

Webster, M., Breza, M., Dixon, C., Fisher, M. and McCann, J. 2020. Exploring the effects of environmental conditions and design choices on IoT systems using formal methods. Journal of Computational Science, 45: 101183.

Wu, Q., Han, J., Cai, C., Huang, H., Gao, S., Zhu, L. and Liu, H. 2022. Formal verification method considering electric vehicles and data centers participating in distribution network planning. In 2022 IEEE 6th Conference on Energy Internet and Energy System Integration (ei2), pp. 3030–3034.

Wu, Y., Hu, M., Liao, M., Liu, F. and Xu, C. 2021. Risk assessment of renewable energy-based island microgrid using the hflts-cloud model method. Journal of Cleaner Production, 284: 125362.

Zhou, S., Qin, L., Ruan, J., Wang, J., Liu, H., Tang, X., . . . Liu, K. 2023. An AI-based power reserve control strategy for photovoltaic power generation systems participating in frequency regulation of microgrids. Electronics, 12(9): 2075.

CHAPTER 10

Direct Current Microgrids:
A Business Model Perspective

Christine Mwase

1. Introduction

Rising global energy demand amidst depleting fossil fuel reserves, higher incidence of climatic shocks, ageing grid installations, and technological advances are motivating a shift from the traditional centralized power grids which have dominated power production and distribution to a more distributed system of energy supply and distribution, in the form of microgrids. The socioeconomic benefits are high, particularly in communities that are in remote and off-grid locations; regions that are highly vulnerable to climatic shocks; and/or where disruptions to electricity could have adverse effects (e.g., hospitals). Disruptions to connections to centralized power grids in some cases lead to near or complete loss of access to powerlines connected to the national grid. By powering the increasing range of applications that utilize DC loads, such as electric vehicles and data centres, directly from distributed renewable energy generators, DC microgrids provide a means to more effectively and sustainably bridge the electricity demand gap.

While the precise definition of microgrids tends to differ (Mauger, 2022; Olivares et al., 2014), in this chapter a microgrid is regarded as a group of interconnected loads and distributed energy resources (DERs) within clearly defined

Fudan University, Shanghai, China.
Email: cmwase@ieee.org electrical boundaries that act as a single controllable entity with respect to the grid. A microgrid can connect and disconnect from the grid to enable it to operate in both grid-connected or island mode (Ton and Smith, 2012). When the loads and DERs are connected to a common direct current (DC) bus, the microgrid is referred to as a DC microgrid. This is in contrast to alternating current (AC) microgrids which use a common AC bus, or hybrid microgrids which use both AC and DC buses.

We focus on DC microgrids due to the growing adoption of direct DC DERs, like photovoltaic (PV) generation, stationary and mobile batteries, and fuel cells —benefits that make DC microgrids crucial for future electricity production. There is also a rising interest in high-efficiency direct DC loads. DC microgrids can eliminate multiple power conversion steps associated with AC buses, resulting in improved energy efficiency and cost-effective operation. Table 1 summarizes some of the benefits which make DC microgrids crucial for future electricity production, both in areas with and those without existing grid infrastructure. Specifically:

- In areas without existing grid infrastructure (off-grid), the primary motivation of DC microgrids is to provide access to energy, a first step in the hierarchy shown in Figure 1. By generating power close to the point of consumption, they reduce losses along transmission and distribution lines, providing efficiency gains. DC microgrids are becoming an attractive technological solution for the primary source of electrification for communities and one of the drivers in affordable and flexible energy systems.

- In areas with existing electrical grid infrastructure, the motivation tends to be more varied beginning with increased reliability and resilience (Garip et al., 2022) as shown in Figure 1, by providing facilities to augment grid supply, particularly in areas with unreliable supply. DC microgrids also provide cost-effective expansion of energy supply from non-grid sources (for example, in remote areas subject to climatic shocks or as a replacement to existing old technology), reduction of greenhouse emissions by incorporating clean energy sources, and offer positive effects from more agile electricity production (including electricity distribution that is better aligned to demand). With the decreasing cost of renewable technology such as solar PV, and the increasing efficiency of battery energy storage systems, there is rising demand for on-site decentralised power generation as an addition or alternative to grid supply. Large-scale investors are also increasingly turning towards DERs due to their ability to defer large power system infrastructure investments, mitigate risks associated with the construction of big power plants, reduce power losses, and improve power quality and energy efficiency.

The development of business models and ownership structures that permit cost-effective operation and the subsequent acceleration of shift towards decentralisation of ownership trend will be crucial to accelerating the adoption of DC

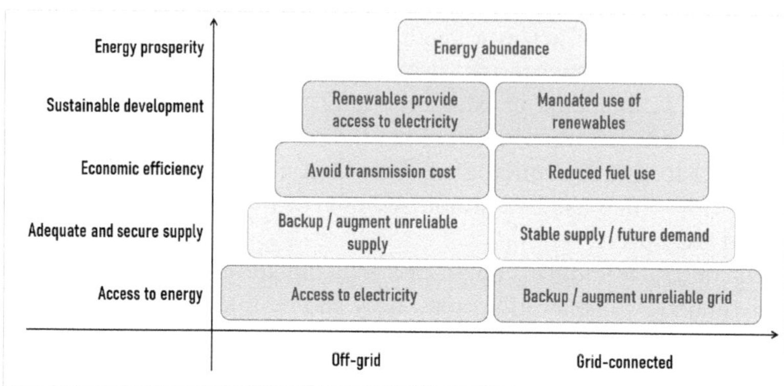

Figure 1. The evolving motivations for DC microgrids – off-grid vs. grid-connected.

microgrids. DC microgrids have yet to achieve significant scale and there is scope for further business model innovation to bring them more broadly into the market. Among the barriers is the lack of long-standing experience with scalable microgrid financing models that potential investors can rely on (NRE, 2019).

The objective of this chapter is twofold: to outline the motivation for the innovation of business models for DC microgrids; and examine plausible business model options that can be used to address emerging gaps in the coming decade.

The remainder of this chapter is organised as follows. Section 2 provides an overview of the key drivers and factors influencing the energy transition, highlighting the areas of disruption that demand innovation. Section 3 discusses potential DC microgrid business models based on existing and emerging trends. Section 4 examines different deployments and assesses their business models. Section 5 collects some perspectives on the future evolution of business models for DC microgrids. Finally, Section 6 draws out the main conclusions and key takeaways.

2. Drivers of the energy transition

2.1. Historical background

DC microgrids stretch back to the origins of the grid in the late 19th century when electrical distribution systems used direct electric current to provide lighting to communities. The adoption of electrical lights resulted in small-scale decentralised power generation and transmission in cities and countries. One such example is Thomas Edison's electric utility company which in 1882 provided DC-based electric lighting solutions. Electricity was generated at one central point in the middle of a city and then transmitted along underground conductors to the homes and offices of users. The high cost of copper wire and high transmission losses however limited the distance that customers could be from

Table 1. Potential benefits of DC microgrids.

	System	Consumer	Society
Autonomy	– Enhances the ability to balance supply and demand without relying on the grid	– Promotes self-reliance and allows for decision-making about energy supply	– Promotes community energy independence – Enables community involvement in energy supply
Cost	– Reduces investment risk and enhances the flexibility of investments – Attracts private investment, helping to spur innovation in energy products and services – Encourages third-party investment in the local grid and power supply – Aids in improving economies of scale	– Reduces energy service costs through decreased purchasing – Increases provider competition and thus increases consumer choice – Facilitates the sale of generated electricity – Reduces price volatility	– Increases economic development and supports employment – Increases efficiency and consumer participation in electricity markets, reducing price volatility and fuel price pressure – Reduces total land-use needs for energy systems
Efficiency	– Energy goals can be optimized around economic and environmental factors	– Promotes customer participation in energy efficiency, demand management and load levelling	– Reduces emissions of pollutants and greenhouse gases
Flexibility	– Facilitates technology neutrality by enabling selection across diverse generation sources – Provides multiple infrastructure construction and operation options – Flexible system expansion given shorter lead times and incremental build-out	– Flexibility to choose the generation sources and infrastructure options that best suit their requirements and preferences, ensuring optimal efficiency and cost-effectiveness	– Flexibility to choose renewable energy sources reduces reliance on fossil fuels, lowers carbon footprint, and helps mitigate climate change.
Reliability	– Enhances DER integration, handling sensitive loads and the variability of the renewables locally – Improves bulk-power-system reliability by enabling multiple loads to be islanded – Reduces likelihood of disruption by placing generation near the consumer – Creates and maintains self-healing networks	– Reduces the number and duration of outages – Ensures highest-reliability energy supply to critical loads – Provides improved power quality, controlled at the local level	– Provides facilities during power outages, ensuring critical services and infrastructure remain operational during emergencies
Scalability	– Reduces system peak loads, promoting demand-side management and load levelling and providing congestion relief – Enables scalability of production and consumption through parallel and modular generation	– Increases energy access	– Enhances energy security

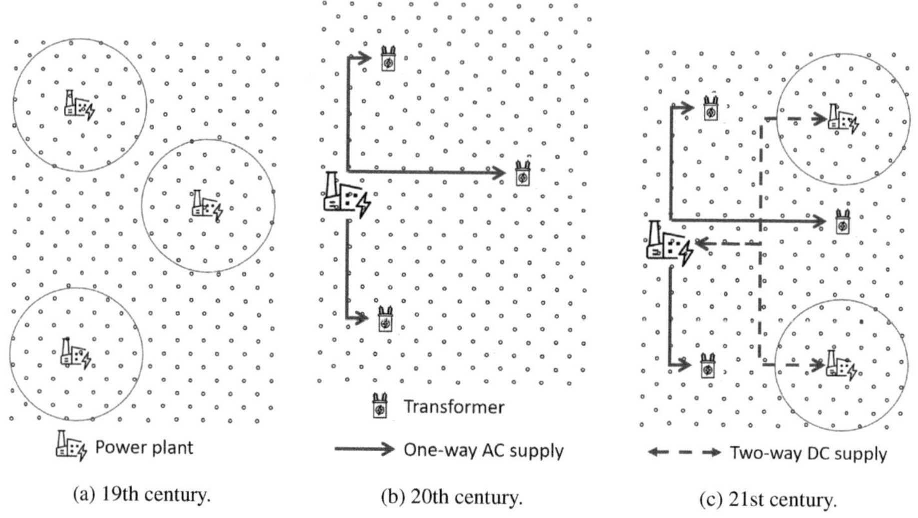

Figure 2. Early stages of the grid's development.

the generation station and consequently, the service area that the systems could cover. This resulted in multiple isolated systems as depicted in Figure 2a, and limited the exploitation of economies of scale, eventually leading to higher costs.

The ensuing period (from about 1920s to 1970s) was marked by a massive wave of electrification, driven by growing demand and bolstered by government support, which centred on centralized generation and transmission taking advantage of economies of scale. Electrical energy was distributed in the form of AC, with AC voltage being raised or reduced using a transformer, enabling efficient long-range power distribution through high-voltage transmission lines as shown in Figure 2b. The amount of power lost during transmission, P_{loss}, is directly related to the square of the current flowing through the wires: $P_{loss} = I^2 R$ where I is the current and R the resistance. By increasing the voltage, the current is decreased proportionally, allowing for the same amount of power to be transmitted with less copper wire and lower energy losses. AC transmission was thus preferred over DC at the time due to its cost-effectiveness. From the 1920s through to the 1970s, the increased reliability afforded by connecting multiple generating units to diverse loads, lower construction costs per kilowatt (kW), and the ability to draw power from distant large generating resources (e.g., large scale hydropower plants) propelled the development of the centralized grids we see today.

A century later, the grid - which benefited from centralised AC generation - is now witnessing the beginnings of a return to decentralisation and an increased use of DC.

- The microgrid concept resurfaced in the 1990s as a solution for incorporating various DERs while also increasing the resilience and dependability of power supply in the event of emergencies and natural catastrophes. The surge was also aided by greater awareness and concern over climate change.

- Since their resurfacing, the benefits of microgrids continue to be appealing while the earlier advantages of AC macro grids appear to have peaked and are being undermined by greater prevalence of climatic shocks which adversely affect access to centralized grids. Electricity generation companies have steadily transitioned to smaller, decentralised units over time, driven by utility restructuring, improved DER technologies, and the economic risks associated with the construction of massive power generation plants and transmission infrastructure (Hirsch et al., 2018). Furthermore, AC distribution is no longer compatible with our increasingly DC world with emerging technologies like battery storage, solar PV for generation, and energy-efficient DC loads.
- Moreover, AC is no longer the only method of transferring power, and DC is used at high power transmission powers. At the start of electric power transmission, the transmission of high voltage direct current (HVDC) was unattainable. There was no cost-effective method to step down the DC voltage for user applications. However, in recent years, HVDC has made tremendous progress in recognising new ways to modify DC voltages.

2.2. Drivers of the energy transition

Three factors are primarily driving the shift from a traditional centralized AC-based grid structure to one with a larger role for DC microgrids.

2.2.1. Energy demands

Firstly, the volume and range of applications requiring electricity have expanded significantly. In recent decades, the worldwide use of energy has risen exponentially, ranging from 8.6 billion tonnes (Btoe) in 1995 to 13.1 Btoe in 2015 (Dong et al., 2020), this being more than 190 qBtu higher than 1990 levels. IEA CPS projections forecast global energy consumption will continue to grow by approximately 20 to 30 percent reaching about 767 qBTU by 2040, an increase of 41% over 2015 (Newell et al., 2019). This constantly rising energy demand is depicted in Figure 3 which presents previous and predicted usage in the coming decades. Traditional energy production is unable to meet this growing energy demand in its present mode (Sadekin et al., 2019), both due to the growing volumes of fossil fuels that would be needed, and the added pressure on ageing grid production and transmission infrastructure. Decentralized generation, bringing in renewable energy, is thus a welcome opportunity.

Secondly, concerted policy efforts to reduce global greenhouse gas emissions while also promoting decentralization of the grid through more autonomous production from DERs are contributing to a pick-up in usage. With the growing

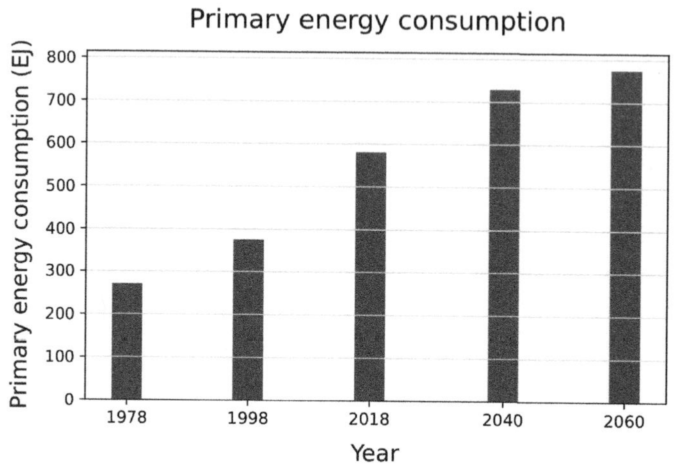

Figure 3. Primary energy demand.

opposition to new fossil-fuelled power stations, renewable generation is on the rise, with a 41–60% increase in the share of renewable energy in total electricity generation expected by 2060 (Kober et al., 2020). This shift to decentralised renewable generation favours modularity over the benefits of scale that AC macro grids offered.

Thirdly, more electronic devices are using DC, and more DC sources are being used for generation. In data centres, for example, almost all of the critical payloads in the data centre are DC loads, thereby reducing the number of state changes and the amount of heat energy wasted. The shift to DC results in gains in availability and efficiency, as well as savings in floor space, cooling costs and overall upfront capital costs. Similarly, native DC equipment (LED lights, mechanical components, etc.) comprises over 50% of a building's total energy load (Glasgo et al., 2016). The shift to DC in commercial buildings thus results in fewer converters and less conversion loss, presenting an opportunity to save energy, money and greenhouse gases generated from our electrical grid. The same is found in aircraft, where hydraulics which are being replaced by electrical actuators make use of smaller converters resulting in weight reduction. These, and other applications, are adding significant additional demands. There is, for example, an expected explosion in electric vehicles (EVs), with 750 million expected globally by 2030 and nearly 3 billion by 2050 – a twenty-fold increase from current levels (Brown et al., 2019).

2.2.2. Technological innovation

Technology to support DC Microgrids has evolved, the most notable being power conversion. Historically, power conversion has been the domain of inverters (DC to AC converters) or rectifiers (AC to DC converters). With the rapid evolution of the DC Microgrid, the DC:DC converter has stepped into the fore of power conversion to serve as the bridge between DC sources and loads interacting in a native DC environment. The advances in power electronics for DC systems result in the lowest number of power conversion steps, thus minimising waste and reducing costs. By leveraging the inherent DC functionality of most DERs, including solar PV and fuel cells, and many end-loads including lighting, power electronics and variable speed drives for heating, ventilation, and air conditioning (HVAC), all-DC microgrids avoid conversion losses which can waste between 5% and 15% of power generation (Hirsch et al., 2018).

2.2.3. Economics

The cost of DC generation and storage, such as solar photovoltaics (PV) and electric storage, as well as power conversion, have fallen precipitously through technological innovation. Within one decade, the price of electricity from utility-scale solar photovoltaics has declined by almost 90% as shown in Figure 4. Unlike fossil fuels and nuclear power whose costs largely depend on the price of the fuel that they burn and the power plant's operating costs, the cost of renewable power is largely dependent on the cost of the power plant i.e., the cost of the technology itself, and decreases with increasing capacity. At each doubling of installed solar capacity, the price of solar electricity fell by 36% (Roser, 2020). On the other hand, despite having the greatest installed capacity, coal has not become significantly cheaper.

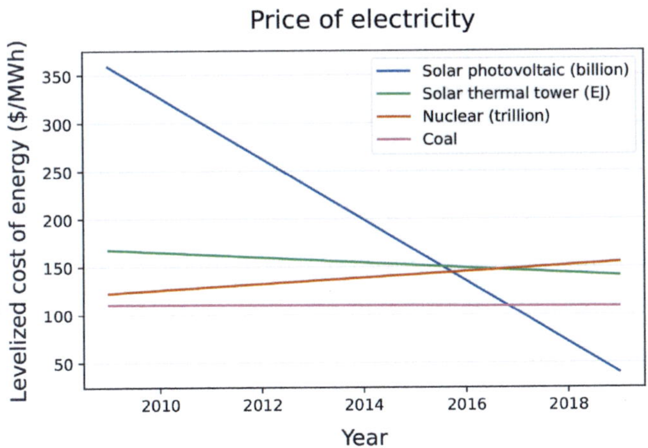

Figure 4. Price of electricity (Roser, 2020).

Rising rates of traditional grid service electricity have also played a role. For decades, the cost to deliver energy through traditional systems has been increasing significantly since the 1990s while the cost of DC DERs such as solar and batteries has been and is expected to continue to decline. Considering increasing energy demands, rising fuel costs, depleting fossil fuels, declining renewable energy costs and ageing infrastructure, the possibility of deferring large power system infrastructure investments through DERs is attractive to large-scale investors and boosting microgrid adoption. In parallel with this, technological advances are enabling the integration of digitization and distributed energy to unlock novel ways of monetizing energy infrastructure and services in ways that were not feasible before.

2.3. Key changes in the value chain

Table 2 lists some of the key shifts in the value chain that are driving disruptive innovation in the industry.

Table 2. Differences between the traditional macro grid and DC microgrids.

	Traditional AC Macro Grid	DC Microgrids
Energy production	Centralized, fossil-based, mostly non-renewable energy sources	Distributed, focus on renewable energy sources
Main participants	Producers and consumers	Prosumers
Digitisation	Little use of ICTs, scarce intelligence	Widespread use of ICTs and intelligence in decision-making
Data	One-way stream, scarce, offline	Two-way interchange, abundant, online
Energy agents	Few agents	Numerous agents

2.3.1. Distributed generation

Energy production has evolved from self-production and consumption to localized production, then to centralized production, and is now undergoing a transformative shift towards distributed generation.

- Prior to utility-centric business models, customers generated electricity on-site for their own consumption, as shown in Figure 5(a), buying equipment and services from equipment providers using a primarily transactional relationship. Utilities then emerged, widening access to more customers by taking care of production, which some potential customers may not have had the space and/or funds to accommodate. The sale of electricity rather than equipment was maintained as utilities shifted from localized production shown in Figure 5(b) to centralized production shown in Figure 5(c).
- The growth of distributed energy is driven by multiple factors. Firstly, there has been strong support from industrial and environmental policies to increase the share of renewable electricity capacity. Initiatives like net metering and feed-in tariffs have provided incentives for individuals to

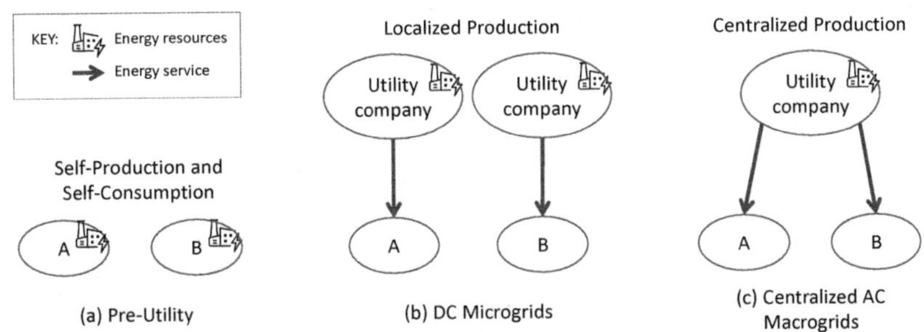

Figure 5. Evolution in production and consumption of electricity. [A, B: customers].

generate electricity. Rising electricity prices have also made it more appealing for consumers to generate their own electricity. This not only helps reduce their reliance on the grid but also allows them to sell excess energy to the grid at higher prices. In addition, the cost of photovoltaic technology and energy storage has significantly decreased, making it more accessible to the masses. Moreover, small-scale distributed generation is an attractive solution in places where grid infrastructure is not in place.

- With distributed generation comes a range of significant characteristics that not only pose new challenges but also open doors for novel business opportunities. Unlike before, electrical energy is now produced at many more spatially distributed nodes and by many individuals and organisations. Integrating distributed generation, especially when it is at the small-scale individual level, into existing markets is a challenge. One of the challenges it brings is the need to reduce intermediary costs in energy trade. Intermediary costs include those associated with metering, billing, administration fees, IT services, banking services and brokerage. Moreover, distributed generation (such as is the case with solar PV), frequently involves intermittent renewables. This has given rise to exciting opportunities such as Peer-to-Peer (P2P) energy trading, which offers a practical way to incorporate intermittent small-scale generation into the system with minimal intermediary costs.

2.3.2. *The emergence of prosumers*

Prosumers are individuals or entities that both consume and produce electricity. In the traditional AC macro grid, consumers were solely passive recipients of electricity. With the introduction of AC technology came long-distance transmission and the traditional value chain shown in Figure 6. Utilities dominated the value chain, generating and distributing energy to customers. Their ownership and authority over the grid infrastructure allowed them to succeed using resource-driven business models, without much effort placed on acquiring and maintaining customers.

Figure 6. Traditional value chain.

However, in DC microgrids, prosumers actively participate in the energy market by generating their own power and feeding energy back into the grid. This shift empowers individuals and organisations to take control of their energy production, promoting energy independence and reducing reliance on centralized power providers. Prosumers also have the potential to participate in peer-to-peer energy trading, further democratizing the energy sector and fostering a more decentralized and inclusive energy market.

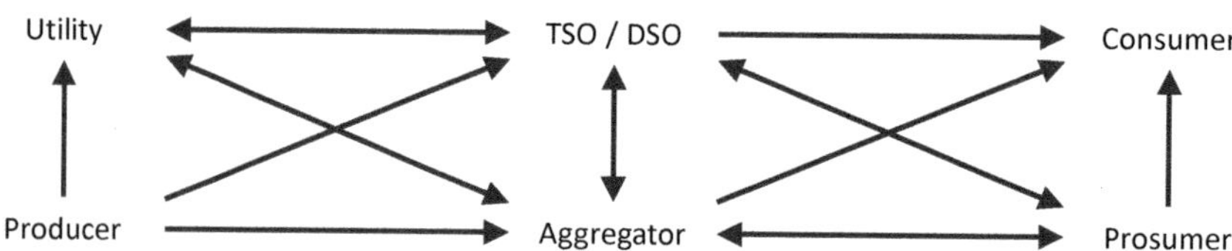

Figure 7. Emerging value chain.

The re-emergence of DC microgrids and the emergence of prosumers have triggered a significant shift from utility-centric value chains and business models to customer-centric ones. Figure 7 shows a simplified version of the emerging value chain. Traditionally, utility companies held a monopoly over energy generation, distribution, and pricing, wielding immense power and control. However, the rapid proliferation of customer-owned and controlled DC microgrids has upended this dynamic. Customers now possess the ability to generate and consume their own power, making them active participants in the energy market. Prosumers have become increasingly valuable to the energy market as they can contribute additional energy to meet the ever-growing demand, thus fostering a mutually beneficial relationship between them and utility companies. Moreover, prosumers introduce their needs, stakeholder relationships, contributions and influence into the grid ecosystem, compelling other players, including utility companies to adopt a customer-centric approach to remain competitive in this evolving landscape. Recognizing that prosumer needs vastly differ from those of utility companies, energy providers are shifting their focus towards meeting and anticipating these unique needs.

2.3.3. *Digitisation of infrastructure*

Digitalization is transforming every sector, including the energy sector, where data can be converted into a valuable resource. While digitization of infrastructure is a key change that occurs in both the AC macro grid and DC microgrid value chains, its importance is more pronounced in DC microgrids due to their complex and dynamic nature. The digitization of DC microgrid infrastructure and the availability of data have significant implications for the way energy is perceived and utilized, ultimately leading to its de-commoditization. Traditionally, energy has been treated as a commodity, with a clear buyer-seller relationship and a limited focus on its source and distribution. However, advancements in technology and the digitalization of energy infrastructure are rapidly shifting this paradigm. The digitalization of DC microgrid infrastructure allows us to collect vast amounts of data related to energy generation, consumption, and distribution.

Figure 8 illustrates the emerging value chain planes resulting from the digitization of the energy sector. The emerging value chain depicted in Figure 7 lies within the electricity plane of Figure 8. With the physical infrastructure that constitutes the grid lies a network from which data is generated and captured. This data offers valuable insights into the dynamics of energy systems, enabling more efficient energy management and optimization. Real-time monitoring and analysis of energy data empower operators to identify trends, patterns, and inefficiencies, leading to more informed decision-making. This transparency and accountability in the energy sector provide consumers with detailed information about their energy usage, including the sources of energy and their environmental impact. Armed with this knowledge, consumers can make informed choices regarding their energy consumption, driving more sustainable and environmentally friendly practices.

Furthermore, the decentralization of energy assets through digitization allows for greater local control and ownership, breaking away from the traditional centralized energy model. This shift promotes energy independence, resilience, and reliability by reducing dependence on large, centralized power plants and transmission networks. As energy becomes more localized, transparent, and opti-

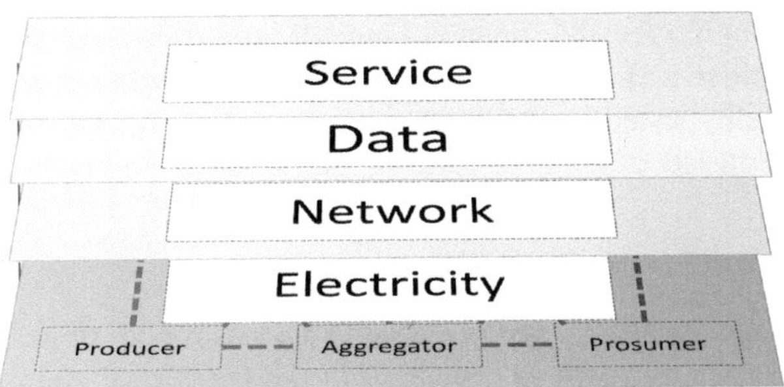

Figure 8. Emerging value chain planes.

mized through digitization and data availability, new business models and market structures emerge. The de-commoditization of energy transforms it from a standardized product bought and sold at market prices to a personalized, value-driven service that caters to specific needs and preferences. This evolution reshapes the energy landscape, moving away from a traditional commodity-driven approach towards a more customer-centric, service-oriented approach.

These rapidly evolving shifts in the value chain demand a paradigm shift in business models for DC microgrids, as traditional approaches fall short in harnessing the full potential of the changing landscape.

3. Business models

The shift from a traditionally centralized AC-based macro grid structure to one with a larger role for DC microgrids brings with it changes in product/service offerings and business models. While the past century was dominated by utility-based ownership and operation, operationalising the grid of the next century requires the involvement of multifarious ownership and management structures. This section looks at the trend away from utility-centric models and presents emerging business models for DC microgrids.

3.1. Definition of business model

The business model concept which dates back to the mid-20th century gained attention with the emergence of internet-based businesses in the mid-1990s. From communicating complex business ideas to potential investors within a short timeframe (Zott et al., 2011), the business model concept has evolved into a tool for systematic analysis, planning and communication, as well as a strategic asset for competitive advantage and firm performance (Geissdoerfer et al., 2020). Despite growing in use, there is no consensus on the definition of the business model. Table 3 provides some selected definitions. We place focus on the following common objectives that most address; (i) the creation of value (e.g., a superior and competitive product or service); (2) the design of a set of components and activities to deliver that value; and, (3) the capturing of value (e.g., the recovery of costs through a feasible financial model), and utilise the business model canvas (Osterwalder and Pigneur, 2010) because it provides a framework that has been thoroughly tested and utilised in the energy sector. The business model canvas (BMC) consists of the nine building blocks described in Table 4.

3.2. The traditional business model

Figure 9 shows the traditional utility-centric business model stemming from the traditional value chain shown in Figure 6. In the traditional energy landscape, utilities held a dominant position, generating and distributing energy to customers using resource-driven business models.

Table 3. Selected business model definitions.

Author	Definition
Timmers, 1998 (Timmers, 1998)	A business model: (i) an architecture for the product, service and information flows, including a description of the various business actors and their roles; and (ii) a description of the potential benefits for the various business actors; and (iii) a description of the sources of revenues.
Afua and Tucci, 2003 (Afuah and Tucci, 2003)	A business model is the method by which a firm builds and uses its resources to offer its customers better value than its competitors and to make money doing so.
Mitchell and Coles, 2003 (Mitchell and Coles, 2003)	A business model comprises the combined elements of "who", "what", "when", "why", "where", "how" and "how much" involved in providing customers and end users with products and services.
Osterwalder, 2004 (Osterwalder, 2004)	A business model is a conceptual tool that contains a set of elements and their relationships and allows expressing a company's logic of earning money. It is a description of the value a company offers to one or several segments of customers and the architecture of the firm and its network of partners for creating, marketing and delivering this value and relationship capital, in order to generate profitable and sustainable revenue streams.
Chesbrough, 2007 (Chesbrough, 2007)	A business model performs two important functions: value creation and value capture. First, it defines a series of activities, from procuring raw materials to satisfying the final consumer, which will yield a new product or service in such a way that there is net value created throughout the various activities. Second, a business model captures value from a portion of those activities for the firm developing and operating it.

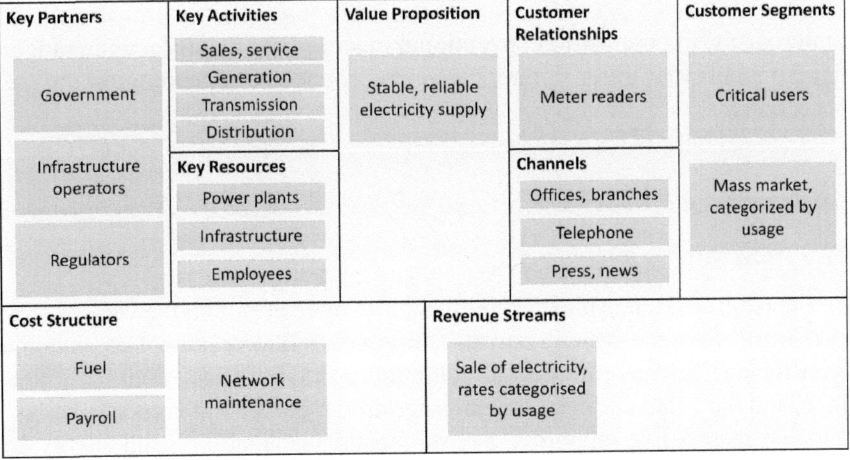

Figure 9. Traditional business model.

Table 4. Business model canvas building blocks.

Pillar	Building block	Description
Product/ Service	Value Propositions	The products and services that create value for the customer
Customer Interface	Customer Segments	The segments of customers a company wants to create value for
	Channels	Describes the various ways a company interacts with its customers to deliver value
	Customer Relationships	Explains the kind of relationships a company establishes and maintains with each of the customer segments, and how it maintains them
Infrastructure Management	Key Activities	Describes the arrangement of key activities required to offer and deliver value
	Key Resources	Outlines the key resources that are indispensable in the company's business model
	Key Partners	Shows the network of partnerships with other companies that are necessary to efficiently offer and deliver value
Financial Aspects	Cost Structure	Represents the sum of costs that the company can or will have for the applied business model
	Revenue Streams	Explains the method a company generates money through various revenue flows which result from successfully delivering the value proposition to customers

3.3. Emerging business models

DC microgrids, in contrast to conventional macro grids, exhibit a wide range of business models that differ within and across countries, depending on the ownership structure.

3.3.1. Conceptual framework

3.3.1.1 Ownership structures

The ownership structures play a significant role in determining the various stakeholders involved in the system, and the subsequent value streams. The stakeholders can include private enterprises, community organizations, and government entities, and the value derived from microgrids can go beyond monetary terms to encompass factors like reliability, environmental sustainability, and energy efficiency. This translates into value streams being measured not only in monetary terms but also in performance metrics, such as energy savings. The diverse ownership structures of microgrids enable innovative business models tailored to the distinct needs and goals of different communities and regions. Figure 10 presents the relationships between the main participants based on the emerging ownership structures.

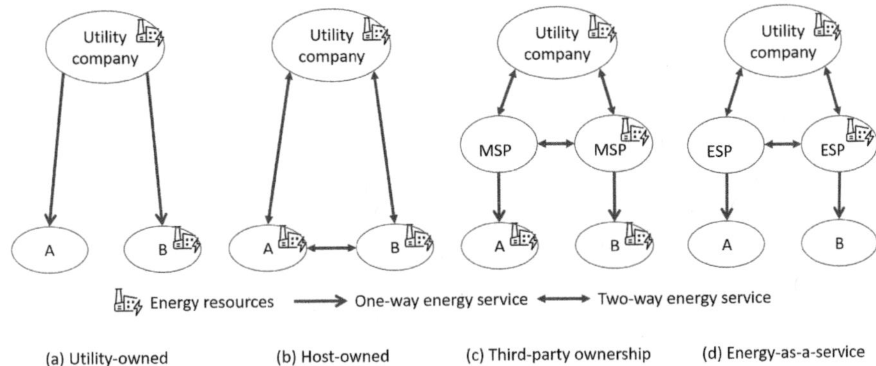

Figure 10. Main actors and relationships in emerging ownership structures. MSP: Microgrid service provider; ESP: Energy service provider; A, B: customers

3.3.1.2 *Contracting vehicles*

An important element is determining the contracting vehicle that best lends itself to the business model and asset ownership taking into account capital investment and operations and maintenance goals. Table 5 summarises the ownership, investment, and operational responsibilities of different business models, mapping them to contracting vehicles. More generally, ownership-based contracting vehicles such as build-own-operate (BOO) and build-own-operate-transfer (BOOT) offer a more predictable revenue stream for the owner of the DC microgrid, allowing for long-term planning and investment albeit at a typically larger upfront investment from the owner. On the other hand, operational-based contracting vehicles such as PPA and energy service agreement (ESA) provide a flexible and scalable energy solution for the client, allowing for a lower upfront investment from the client.

3.3.1.3 *Evaluation metrics*

The business model can have a significant impact on the nature and weight given to different evaluation metrics. Some commonly used evaluation metrics in traditional business models are applicable, but the weight given to each metric will depend on the ownership model. Table 6 maps business models to commonly used evaluation metrics.

- Profitability metrics such as ROI, net present value (NPV) and IRR compare the expected benefits (e.g., revenues, cost savings) to the initial investment, providing an assessment of the associated profitability and risk, such as when large upfront costs are made and maximising financial returns are a priority. On the other hand, cost metrics such as COE and

Table 5.　Summary of business models and contracting vehicles.

Model	Asset Ownership	Capital Investment	Operations and Maintenance	Contracting Vehicles			
				BOO/BOOT	LTO	PPA/ESA	JDA
Utility owned	Utility company	Utility company	Utility company*	X		X	
Host owned	Host entity	Host entity	Host entity*	X		X	
Third-party owned	One or several entities**	Third party	Third party	X	X	X	
Energy-as-a-service	Investor(s)	Investor(s)	Service provider			X	
Mixed ownership	Multiple entities	Investor(s)	Service provider			X	X

BOO = build-own-operate; BOOT = build-own-operate-transfer; LTO = lease-to-own; PPA = power purchase agreement; ESA = energy service agreement; JDA = joint developer agreement

*Can be in-house or outsourced to a third-party service provider or contractor.

** Can be transferred to an off taker.

Table 6.　Mapping of business models to metrics.

Model	Metrics			
	Profitability ROI/NPV/IRR/PP	Cost COE/LCOE	Risk PD/EL	Operational CF/Availability
Utility owned	✓	✓		
Host owned	✓	✓	✓	
Third-party owned	✓	✓	✓	
Energy-as-a-service		✓		✓
Mixed ownership				

levelized COE (LCOE) provide a different perspective, focussing on cost-effectiveness, such as where minimizing energy production costs is a primary concern. Risk metrics, such as probability of default (PD) and expected loss (EL), can also be used to assess the potential risks, while operational metrics, such as capacity factor (CF) and plant availability, can provide insights into the performance and reliability.

Figure 11 presents multiple dimensions to be considered when determining evaluation metrics. While the BMC traditionally considers the economic dimension, within which customer-, offer-, finance-, and resource-driven BMCs can be developed, recent studies emphasize the importance of environmental and social

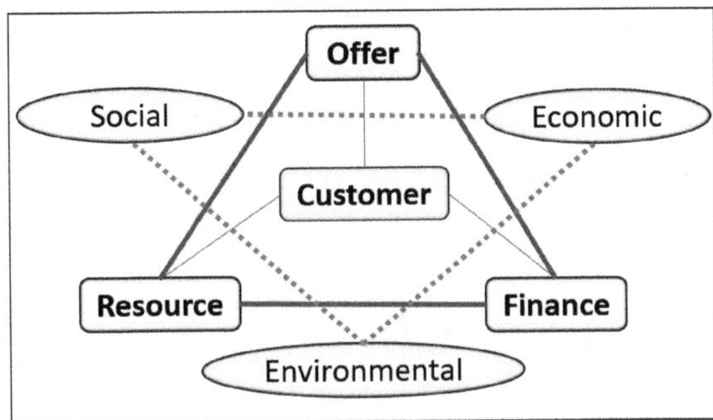

Figure 11. MultiBMI framework.

dimensions as well. Each of these dimensions can be further broken down into specific metrics that can be used for assessing performance and impact.

3.3.2. *Implications on business models*

This subsection summarises the main business models based on the emerging ownership structures of DC microgrids.

3.3.3. *Utility-owned*

While microgrids were once considered a threat to utility companies' revenue and balanced operation, technological innovation and cost reduction of DERs have made them valuable additions to utilities from both a supply and demand perspective. Specifically:

- From a supply perspective, utility companies typically find it beneficial to own DC microgrids when building a microgrid is cheaper than building a parallel transmission line and ensuring the required reliability level (such as in remote areas). It is also applicable, as highlighted by the customer segments in Table 7, in areas where reliability through the traditional grid has proved difficult, such as in areas that experience recurring flooding and subsequent outages. When outages are not expected, additional revenue can be earned by arbitraging from cheap to expensive generation.
- From a demand perspective, the customer base for utility-owned micro-grids can come from industries and businesses seeking clean energy and/or highly reliable energy sources without the burden of managing their grid. The utility company would install the microgrid at the customer premises as shown in Figure 10(a) and assume all duties associated with its operation.

By building DC microgrids, utility companies defer transmission and distribution costs, achieving less costly service expansion and increased grid reliability. Furthermore, the surplus energy generated is additional energy for the grid and

Table 7. Business model options.

Model	Description	Value Proposition	Customer Segment	Infrastructure	Revenue Aspects
Utility owned	Utility company owns microgrid facilities. Microgrid operations prioritize the objectives of the bulk system.	- Increased grid resilience and reliability - Shorter waiting times for new installations - Increased renewables in the energy mix - Value-added services	- Remote communities - Outage-prone locations - Clean energy seekers - Essential service providers (e.g., hospitals, military bases, airports)	- Energy generation and storage assets (e.g., solar installation, batteries) - Maintenance and operation of generation and distribution systems - Connections to producers of DERs and microgrid components and external technicians	- Fixed price per kilowatt-hour - Revenue from arbitraging
Host owned	Microgrid facilities are owned by the host entity.	- Reduced electricity bills - Revenue from selling electricity - Government subsidies	- Self - Utility companies - Community members	- Distributed generation assets (e.g., solar installation) - Links to manufacturers of DERs, installation companies, financiers and utility companies	- Revenue from selling energy and associated benefits
Third-party owned	Microgrid facilities are owned and operated by one or several entities.	- Provide electricity at lower, predictable costs - Grid reliability - Secure energy supply - No upfront costs to microgrid user - No operation and maintenance	- Households, SMEs, industrial and commercial enterprises - Utility companies	- Deployment of microgrid - Delivery of energy services - Securing available subsidies (e.g., tax credits, renewables subsidies) - Networking with government agencies, utilities and other energy/ microgrid sector firms	- Revenue from sales through service contracts or PPAs
Energy-as-a-service	Microgrid facilities are owned and operated by the service provider.	- No upfront capital costs to the user - Predictable energy costs - Ensured reliable energy supply	- Households, SMEs, industrial and commercial enterprises	- Energy and customer management systems - Partnerships with local utilities and energy suppliers	- Revenue from service fees or contracts
Mixed ownership	Microgrid ownership is shared by multiple entities	- Shared investment cost and risk - Diverse expertise and resources	- Households, SMEs, industrial and commercial enterprises - Utility companies	- Deployment and delivery of energy services - Management of skateholder and partner resources and relations	- Revenue from sales through service contracts or PPAs

increases the renewables in the energy mix. Ownership-based contracting vehicles such as BOO, shown in Table 5, can be adopted. Compared to other business models whose financing can be challenging, utility-owned microgrids can often be funded by including the capital cost in the utility's rate base, provided the utility can demonstrate the need for and cost-effectiveness of the microgrid to its regulators.

3.3.4. Host-owned

Individuals or organisations can play the role of being a consumer as well as a producer, selling surplus energy beyond their consumption, as shown in Figure 10 (b). Potential hosts include essential service providers that need to survive when large grid outages occur (such as hospitals, military bases and airports), households that are enthusiastic about environmental protection and have a high payment capacity, small and medium-sized enterprises who have advantages in the development and operation of DERs (such as those encouraged to build their own renewable power system financed by authorities), and individuals who have sufficient and independent roof space (such as farmers) (Cai et al., 2019).

As shown in Table 5, the host assumes full responsibility for the microgrid. This model has a high upfront cost and long payback period and is thus widely used by universities, institutions, military bases and industrial customers and businesses, where large initial investment cost can be overcome. Unlike in the utility-owned model where the utility has the full authority to manage and control microgrid production and maintain grid balance, in the host-owned model the host plays an active role in contributing to grid balance activities.

3.3.5. Third-party ownership (Microgrid-as-a-Service)

While the benefits of microgrid ownership are plenty, the upfront costs, management complexity, and associated risks (e.g., performance risks on an energy efficiency project), can be a barrier to installing microgrids. This steers many potential hosts away from ownership to outsourcing their microgrid projects. In the third-party ownership or Microgrid-as-a-Service model shown in Figure 10 (c), a third party (e.g., solar service company) owns and operates the service (e.g., solar panels installed on the rooftops of consumers' households), while customers pay for the power they consume, according to their service contract or power purchase agreement (PPA). This model has the advantage that customers do not need to pay high installation costs, and do not bear the risk of system operation, overcoming the obstacles of financing and low profitability. The service provider takes responsibility for the installation, maintenance, and performance of energy efficiency measures. In turn, for investors, the key value proposition is the creation of valuable energy services and remuneration streams.

Different contracting vehicles can be considered as shown in Table 5. For example, an energy savings performance contract (ESPC) can be agreed upon between an energy service company (ESCO) and a customer, with the ESCO being responsible for the design, installation, financing, and maintenance of the microgrid, and the customer paying a fixed fee or a share of the energy savings achieved by the microgrid. This model allows customers to avoid the high upfront costs and managerial complexity associated with microgrid ownership while capitalizing on the ESCO's expertise and experience.

A combination of cost and profitability metrics shown in Table 6 may be prioritized, as they are responsible for both generating electricity at a competitive cost and maximizing financial returns for their investors. Table 8 shows one such example where a third-party owned floating photovoltaic (PV) plant connected to the national utility grid 110 kV at Da Mi hydropower reservoir in Binh Thuan province, Vietnam (Nguyen et al., 2023) was analysed considering a plant capacity of 47.5 MW, power output of 69.99 million kWh/year, 20-year PV system lifetime and a total investment of 1,438,877 million VND. For a comprehensive evaluation of profitability, risk, efficiency, and liquidity, net profit value, internal rate of return (IRR), benefit-cost ratio (B/C), and payback period (PP) were used. Furthermore, a sensitivity analysis was used to gauge uncertainty impacts by identifying key metric drivers and assessing assumption changes.

Table 8. Economic-financial indicators of a floating PV plant (Nguyen et al., 2023).

	Economic indicators	Financial targets		Sensitivity analysis			
				A	B	C	A,B
Financial discount factor (iF) (%)		6.95					
Electricity price (US$/kWh)	0.145	0.0935					
IRR (%)	17.2	11.14	IRR (%)	8.43	8.08	10.9	5.70
Net Profit Value (billions VND)	664.3	220.3	Net Profit Value$_f$ (billions VND)	82.9	56.8	205.50	79.2
B/C	1.55	1.211	B/C$_f$	1.073	1.055	1.20	0.931
PP (years)	9.3	14.4	Discounted payback time (yrs)	17.5	18.0	14.7	¿20

A = 10% investment capital increase; B = 10% power generation reduction; C = 10% O&M cost increase.

3.3.6. Energy-as-a-Service

In the Energy-as-a-Service (EaaS) model, a service provider operates the service, and sells energy to customers, typically through a predictable and recurring subscription-based service, as shown in Table 7. In return, the customer is guaranteed a reliable and often elastic supply of energy without assuming the capital investment or management requirements associated with hosting microgrid facilities. This model offers a convenient solution for customers while providing the service provider with stable and recurring revenue streams through the provision of energy services and the opportunity to scale operations over time.

3.3.7. Mixed-ownership

The mixed ownership approach favours private-public partnership (PPP), combining public institutions, utilities, private companies and end consumers. A multi-stakeholder approach decreases investment requirements from an individual stakeholder and can guarantee revenue stream, for instance, based on PPAs. This model can be applied to strong public institutions and high-income communities, who can afford the investment and become shareholders of the microgrids (Vanadzina et al., 2019).

- An Anchor-Business-Community (ABC) collaboration between an anchor institution, such as a large hospital, university, or telecom tower, local businesses and industries, and a community can combine the need for a significant and stable energy demand to support the feasibility of a microgrid project, with the provision of reliable and cost-effective energy for businesses and the community that would otherwise have limited revenue.
- It can also be applied to lower-income communities. For example, a local community may identify its energy needs and prioritise productive uses of energy (PUE) to drive economic development. A microgrid, jointly owned by the community, a private sector investor, and a government agency would allow for shared resources, expertise, and responsibilities. The microgrid would power various productive activities such as agricultural processing and small-scale manufacturing, supporting income generation, job creation, and overall economic growth. Surplus energy can be traded with the main grid, generating additional revenue streams.

Evaluation metrics for mixed ownership models are dependent on the nature of the involved entities. A combination of financial metrics from Table 6 may be required to ensure the feasibility of the project. In additional, evaluation metrics from the environmental and social dimensions shown in Figure 11 may also be required to meet the joint interest of the participating entities. Table 9 shows one such example of additional environmental and socio-economic considerations from a feasibility analysis of installing solar PV panels over California water canals in California (Kahn and Longcore, 2014).

Similar examples can be found in the studies shown in Table 10, which also utilize standard metrics for analyzing energy projects, such as ROI, cost of energy (COE), and life cycle conversion efficiency (LCCE). While the financial metrics are central elements of the revenue aspects highlighted across Table 7, installations in the health care centers and hospitals itemized in Table 10 have needs beyond the economic dimension.

Table 9. Environmental and socio-economic indicators of solar canals (Kahn and Longcore, 2014).

System	Environmental benefits			Environmental costs		Socio-economic benefits	
	Energy saved (MWh/yr)	CO2 abated (MT/yr)	Water Saved (gal/yr)	Manufacturing CO_2 emissions (MT/lifetime)	Cadmium emissions (g/GWh)	Jobs from installation	Jobs from O&M
500 kW	838.8	232	2565463	39.8	0.159	19.9	2.5
1 MW	1677.6	465	5130926	79.7	0.319	37.4	7.5
10 MW	16776.3	4648	51309254	796.9	3.188	373.8	75

Table 10. Recent feasibility studies with economic/financial modelling.

Study and Location	Description	System Specification	Evaluation
Islam et al. (2023) Bangladesh	Powering a rural healthcare center in Bangladesh with a hybrid renewable energy system	Rooftop PV system: 400 Wp solar panels, 25 kW bi-directional inverter, 28 kW generator	ROI: 9.8%; IRR: 12.7% Discounted payback: 6.95 years PP: 7.53 years
Aisa et al. (2022) Libya	Powering Sabratha's general hospital in Libya with a solar-wind hybrid power plant	Solar-wind system: 10 kW wind turbine, 350 W solar panel, batteries, generator	Optimization performance: PV array 1507.73 MWh/yr Wind turbine 596.81 MWh/yr Generator 588.31 MWh/yr COE: $0.182/kWh
Sharma et al. (2019) India	Powering a hospital building in India with a solar PV power plant	Rooftop PV system: 83 PV solar modules, 34.35 kW inverter	PP: 6 years LCCE: 0.007

4. Examples of DC Microgrids

4.1. Case study 1: Energy Local

4.1.1. Motivation

Having seen a five-fold increase in renewable generation over the past decade, the UK's renewable generation efforts shifted towards optimizing and maintaining well-functioning local distribution networks.

4.1.2. Solution

An Energy Local Club (ELC), composed of a local group of customers/households and a local group of renewable energy generators is formed in partnership with a fully licensed supplier. The licensed supplier manages billing,

Table 11.   Energy local (Barnes et al., 2022).

Case	Energy Local
Country:	United Kingdom
Location/scale:	Rural
Established:	2016
Customer segments:	Local renewable generators; local consumers/households
Value proposition:	Better income for the generators; lower bills for households
Infrastructure:	- ELCs are setup as cooperatives, and together with licensed suppliers collaborate to manage a 'complex site', a regulatory exception under Ofgem, the energy regulator. - Smart meters record generation and demand. Virtual meter points measure imports/exports of power not generated/used locally for settlement as the collective outcome of the complex site. - Suppliers have PPAs with generators and modified supply contracts with households.
Revenue aspects:	Energy generated locally is distributed to members under the 'match tariff' and additional demands are met by the licensed supplier through a time-of-use tariff (ToUT).
Socio-economics:	- Increases accessibility to renewable energy for members with limited financial resources or inadequate space for equipment. - Empowers members to be active in energy systems and decarbonization activities. - Strengthens community relationships and fosters social cohesion through clubs. - Supports the achievement of Net Zero emissions by 2050 through efficient and flexible use of existing generation and demand-side assets.

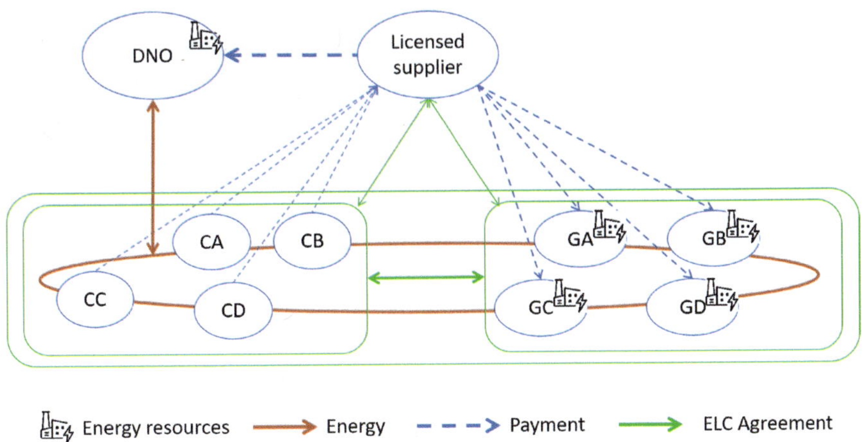

Figure 12. Energy Local model.
CA, ..., CD: consumers; GA, ..., GD: generators

Table 12. Zuiderlicht.

Case	Zuiderlicht
Country:	The Netherlands
Location/scale:	City
Established:	2013
Customer segments:	Residents who do not have access to roof space; roof space owners
Value proposition:	Amsterdam residents can invest in, own, and manage renewable generation assets; offers lower energy bills to members
Infrastructure:	- Energy cooperative projects use the Dutch premium feed-in tariff (SDE+, SDE++) and the Dutch "postcoderoos" regulation subsidy schemes. - Net metering calculates members' electricity imports matched to their share of generated power, and any top-up needs. - The public grid stores electricity not consumed at the time of generation.
Revenue aspects:	- Dutch subsidies for renewable energy provided stable revenue projections, encouraging local community investment.
Socio-economics:	- Increased distributed energy production fosters energy independence and resilience. - Promotes energy literacy through outreach activities that engage and educate residents in renewable generation and the energy transition. - Fosters inclusive participation in the energy transition by allowing individuals of all socioeconomic backgrounds to engage in sustainable energy initiatives.

compliance, and back-office services, including wholesale trading, metering, and customer support. Locally generated energy is shared across ELC demand members using an annually agreed upon "match tariff" that is more competitive than the open market for both groups. For members with greater energy demands, the supplier provides a time-of-use tariff (ToUT) to customers for power that is not generated and matched locally. ELCs are self-sustaining, not through the generation of profit, but rather through the generation of value for multiple actors both within the ELC community and across the wider energy ecosystem. The ELC model, a brainchild of Energy Local (Barnes et al., 2022), is being expanded through Energy Local Community Interest Company (CIC), who acts as the development hub and 'franchise owner'.

4.2. Case study 2: Zuiderlicht

4.2.1. Motivation

A group of local activists created the Zuiderlicht energy cooperative in 2013 with the goal of establishing a more democratic and sustainable energy system (Hansen and Barnes, 2021).

4.2.2. Solution

Zuiderlicht facilitates two types of projects that bring together members with available roof spaces and those seeking to invest in renewable energy.

- Supported by the Renewable Energy Production Incentive Scheme (SDE+), the SDE-backed project allows members to invest in solar PV and gain ownership and management of the system in exchange for their investment. Zuiderlicht negotiates long-term leases with rooftop owners, enabling members to earn a return on their investment. Building tenants benefit from power at rates below market prices, and excess power is sold to a supplier under a PPA.
- Enabled by the Dutch "postcoderoos" regulation (PCR), the PCR-backed project allows members to invest in cooperatively-owned renewable energy generation assets in their local or neighbouring postcode and receive a tax reduction on electricity imports based on their share of generated electricity. Zuiderlicht partners with a supplier who sells generated electricity to members under a modified supply contract and shares the tax discount granted under the postcoderoos regulations between investors and consumers (Hansen and Barnes, 2021).

In contributing towards the goal of running Amsterdam on clean energy by 2025, Zuiderlicht founded Platform 02025 (020) which brings together all energy communities active in the city of Amsterdam.

4.3. Case study 3: sonnenCommunity

4.3.1. Motivation

sonnenCommunity was established to enable energy independence and disrupt traditional energy suppliers by virtually connecting privately-owned storage units.

4.3.2. Solution

The sonnenCommunity platform enables participants to cover 100% of their electricity needs through self-consumption and community energy trading. It operates as a virtual power plant (VPP) and aggregator, linking privately owned generation and storage units via cloud-based software to maximize collective self-consumption and provide ancillary services to the grid. Surplus power from members is deposited into a virtual energy pool, which is then shared among other members. A centralized software platform connects and monitors the members, ensuring a balance between energy supply and demand. This approach eliminates the necessity for a traditional energy supplier (ElMaamoun, 2021; Karami and Madlener, 2018; Koreneff et al., 2020).

The sonnenCommunity operates at multiple levels, including product ownership, asset sharing, and electricity supply contracts. If the community lacks suf-

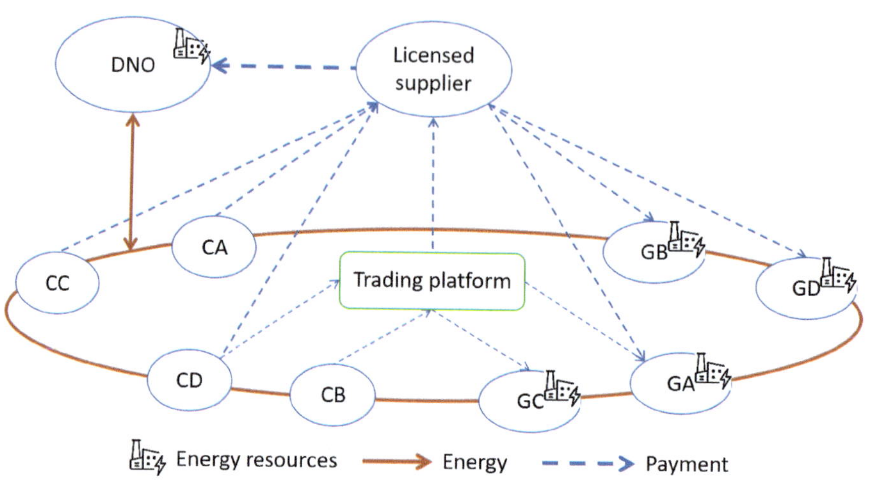

Figure 13. sonnenCommunity model.
CA, ..., CD: consumers; GA, ..., GD: generators

ficient electricity, it purchases from the market or obtains it through PPAs with distributed generators. In these instances, sonnen functions as an energy utility with a new service contract.

4.4. Case study 4: BlockEnergy

4.4.1. Motivation

Utility companies in the United States are faced with the challenge of fulfilling the ever-increasing demand for increased load volume as well as clean and reliable energy. Previously, addressing power consumption growth meant constructing new costly power plants and transmission lines, both decisions carrying substantial long-term consequences. As a subsidiary of a utility provider, Emera Technologies (now BlockEnergy) aimed to tackle the issue of renewable energy while ensuring scalability and compatibility with existing utility systems, providing customers with a simpler and more reliable source of energy, and alleviating them from the complexities associated with its production and distribution.

4.4.2. Solution

The solution that BlockEnergy provides is a utility-focused business model for DERs, which is owned, operated, and maintained by the local utility. BlockEnergy uses a modular platform to connect local on-site PV generation and storage in self-sustaining block loops, creating a mesh network. This network is built in new residential communities, with solar batteries and controllers installed in each home. These homes are further grouped together in a shared network for central storage, which is ultimately connected to the grid. This approach caters specif-

Table 13.   sonnenCommunity.

Case	sonnenCommunity
Country:	Germany
Location/scale:	National
Established:	2016
Customer segments:	sonnen GmbH product or service customers
Value proposition:	Energy independence
Infrastructure:	- Smart programming in each battery optimises individual self-consumption - SonnenVPP software for the VPP operated by sonnen GmbH which acts as a licensed supplier, managing regulatory compliance and providing back-office services. - PPAs with distributed renewable generators for electricity needs not met internally
Revenue aspects:	Flat service contracts promote equity among members. Members benefit financially by allowing sonnen GmbH to control their assets for services to national energy systems.
Socio-economics:	- Increased energy independence through self-consumption and community supply. - Onsite consumption reduces demand on the grid.

ically to new residential construction, allowing for incremental, capital-efficient investments to be made gradually over time. It thus leverages advancements in information and technology availability to make wiser investment decisions. Additionally, it seamlessly integrates with new construction projects without disrupting existing infrastructure or requiring the removal of previously deployed equipment.

Beyond providing distributed energy generation and storage, it also serves as a backup generation and storage solution for the utility. It operates in a more reliable manner since it uses local generation and storage, preventing outages caused by failures in the energy delivery chain. Traditional power grids are interconnected in a series, which means that if an error occurs in the energy delivery chain, it affects the entire system. In contrast, the modular and DC-based design of the BlockEnergy system leads to improved reliability, as most energy generation and storage occurs locally. This also makes it easier to identify and locate issues within the system, resulting in quicker responses and enhanced outage prevention. By positioning BlockEnergy as a capital asset for distributed generation, the solution aims to meet the increasing demand for clean and reliable energy in a more sustainable and cost-effective way (nml, 2021).

4.5. Case 5: Nagoya landfill

4.5.1. Motivation

The motivation for the Nagoya landfill DC microgrid was based on the desire of the landfill owners to enhance the value proposition of their property. They were faced with a paved-over landfill that was unsuitable for most construction purposes. At the same time, as a result of Japan's increasing adoption of distributed renewable energy systems, the local power supply infrastructure had become saturated with solar PV. Japan's objective was to reduce reliance on imported fossil fuels and decommission large centralized nuclear power plants, making utilities and regulators prioritize the consolidation of distributed solar PV. Coupling these challenges with their desire to increase the value proposition of their property, the landfill owners considered the construction of a microgrid with energy storage, enabling them to export dispatchable renewable power and take advantage of a special feed-in tariff.

4.5.2. Solution

In 2014, the Nagoya landfill microgrid facility was constructed at a cost of approximately \$1.5 million (Wang et al., 2020). The facility includes a 0.5MW solar power generation unit and a 0.2MW/1.2MWh battery storage system, covering an area of 2 acres. With its long-lasting battery that harnesses solar PV energy, this microgrid provides a stable power supply to the grid during peak demand periods and serves as a reliable backup for local loads in the event of a grid outage.

The primary investors contributed around 10% of the funding, with the remaining amount secured from a third party through steady payments spanning the project's 15-year lifespan. Special tariffs paid by the utility served as the primary source of revenue. With favourable tariff structures, the payback period for this system was estimated to be around 4 years. Optimal Power Solutions, an Australian firm, spearheaded the development of the facility which was hosted by the landfill owners. Future expansion was envisaged, where several systems like this could be merged into a single virtual power plant (Peter Asmus, 2018).

4.6. Case 6: IIT Madras Uninterrupted Direct Current (UDC) solution to power all homes

4.6.1. Motivation

This case was motivated by the need to not only expand the electricity grid across India, but also to address and improve the issues of affordability, reliability, and overall quality of the power supply.

- About 20% of India's population (approximately 240 million people) had no access to electricity in 2015 making it the largest population in the world without access to electricity (IEA, 2015). Furthermore, a significant proportion of households with electricity endured frequent power disruptions and inferior power quality, with 50% of households remaining electricity-deprived despite being connected to the nationwide grid network.

- While subsidized tariffs were offered in some parts of the country, significant portions of the population were still unable to afford the discounted rates, ultimately leading to financial losses for the power distribution companies. In order to overcome these challenges, it was crucial to reassess the economic barriers in place. The average tariff for homes was approximately ₹5 ($0.07) per unit (considering an exchange rate of US $1 = ₹70). A small home operating two tube lights for six hours, two fans for twelve hours, two bulbs, a 24-inch TV for ten hours, and charging a cell phone for four hours would consume a little over three units of electricity per day. This total cost of around ₹500 per month challenged the ability of at least 50% of homes, particularly those in rural areas, to maintain a consistent electricity supply. Without the incentive to provide power at such rates, power distribution companies would implement load shedding which while being problematic lessened their losses.

4.6.2. *Solution*

The Indian Institute of Technology (IIT) Madras in conjunction with industrial partners developed the Uninterrupted Direct Current (UDC) solution with a view to provide 24 X 7 power to all homes in India.

- Recognizing that DC appliances are more energy efficient than AC ones, the UPC solution innovates on load management by providing a limited but uninterrupted DC power supply to homes (Kaur et al., 2015). Given that PV panels, batteries, consumer electronics, LED lighting, and a growing range of appliances all work with DC, using DC becomes more efficient as it minimizes conversions. The DC system avoids the inefficiency of converting the panel's direct current to AC for synchronizing to the main grid, the conversion back to DC to charge the battery, and a third conversion from DC back to AC when the battery is discharged.
- While India's main power grid is based on AC, their solar-powered microgrids utilize DC power to minimize power losses which can range from 5 to 20% with each conversion between AC and DC. The 125- to 500-watt microgrids serve as a backup power supply for grid-connected households, and as the sole source of electricity for homes not connected to the grid. For grid-connected homes, an additional 48-volt DC power line runs in the home, and the traditional electricity meter is replaced by a UDC power meter which has the same control and communications capabilities of a smart meter, along with an AC-to-DC converter for converting a portion

of the incoming AC to DC. This provides about 10% of the typical household load.

The project was rolled out carefully beginning at a more micro level before spreading out. The field testing of UDC systems commenced in 2014 across a multitude of facilities at IIT Madras, including homes, offices, and dormitories. In 2015, IIT partnered with the Hyderabad-based solar power company Cygni Energy to commercialise UDC systems, extending deployment to 1000 households in three cities and numerous villages. Funding from India's Ministry of Power further facilitated expansive projects aimed at serving over 100,000 households in the semi-urban town of Sasaram, located in Bihar, India. Customers paid for the DC wiring and installation in their households. Thereafter they would be provided with an additional DC power line and would use DC appliances that save more than 50% on power usage (e.g., a DC fan that consumes 30W instead of 72W). For grid-connected households, this would mean savings on electricity bills and continued power supply during load shedding (Jhunjhunwala, 2017).

4.7. Case study 7: Mabushi solar

4.7.1. Motivation

Despite being an economic powerhouse in the African region, access to energy in Nigeria remained a significant challenge with 42% of Nigerians either not connected to the grid or lacking reliable access to energy. This issue was compounded by frequent and prolonged power outages, which result in estimated annual economic losses of $26.2 billion (Ariemu, 2023; World Bank, 2021). As a result, 77% of the country's power demand was being met through self-generation, predominantly through the use of diesel generators. For commercial and industrial customers, roughly 14 GW was generated with diesel generators that are more expensive than the grid. Given these challenges and the growing commitments towards sustainability, there was a rising preference for decentralized microgrid deployments in Nigeria. Not only did microgrids offer a technological and commercially viable solution, but they also aligned with corporate mandates regarding sustainability.

4.7.2. Solution

Mabushi Solar is a project located in Abuja, Nigeria's capital city, for the Federal Ministry of Works and Housing (FMWH). Completed in 2021, it was developed to address multiple challenges that include the lack of reliable power affecting worker productivity at the ministry's Mabushi office block, excessively high electricity bills due to the use of diesel generators as a backup power source, renewable energy and energy efficiency targets set by the Federal Government of Nigeria (FGN), and the aspiration from FGN to explore microgrids as an energy solution and offer a model for other renewable energy developers.

Owned by the federal government through FMWH and financed by the issuance of Green Bonds, the grid-connected plant was constructed by EM-ONE Energy Solutions Ltd. (EM-ONE), a Canadian-Nigerian engineering and technology company. EM-ON worked with FMWH and FGN to address the aforementioned challenges and contribute towards SDGs 7, 12 and 13 through the provision of a 1.52 MWp solar PV array with 2.28 MWh/1.17 MW battery energy storage. In addition to replacing over 400 air conditioners and over 2600 lightings in the Mabushi building complex, an annual reduction of almost $600,000 in energy bills and 2,600 tonnes of CO2 emissions is expected, as well as the addition of 558 new employment opportunities (UN in Nigeria, 2021). The project's goals and anticipated outcomes include an annual production of 2.45 GWh of renewable power, a 78% reduction in diesel consumption, a 70% decrease in non-renewable power usage, a 76% saving on energy expenses, and a 40% reduction in overall energy consumption from retrofitting and efficiency improvements (Premium Times, 2021).

5. Future perspectives

The future outlook for DC microgrids will be significantly shaped by the ongoing trends and developments across various sectors within the energy ecosystem.

5.1. Dynamic boundary microgrids

Dynamic boundary microgrids have the capability to adjust their borders in order to ensure a reliable power supply to critical loads such as hospitals, during expected and unexpected events, in both grid-connected or islanded modes. By introducing flexibility to the boundaries of DC microgrids at different system levels, it becomes possible to maintain efficient and cost-effective operation under normal conditions, while also providing resilience during abnormal situations. This adaptive nature allows for the optimization of energy usage and system stability by expanding or contracting boundaries, thus enhancing the overall resilience of the microgrid system. This flexibility also has the potential to influence the evolution of DC microgrid business models. For example, it can enable the development of more scalable offerings that can cater to a wider range of customers or locations, or incentivize more demand-based offerings that facilitate transactions between microgrids based on real-time fluctuations in demand and supply.

5.2. Virtual power plants

Virtual power plants (VPPs) aggregate energy production, storage and consumption resources, enabling the management of flexible capacity on a large scale for the benefit of their stakeholders.

- The development of VPPs brings about new business models that revolve around DER aggregation, energy trading, demand-side management, provision of ancillary services, and innovative offerings by energy service providers. These flexible portfolios of energy assets that can include DERs with differing characteristics, such as size and application, form individual entities which operate similarly to conventional power plants, making contracts in the wholesale market and offering services. They can for example open up assets to participate in the energy market, generating increased returns for asset owners, or provide grid ancillary services such as frequency regulation and reserve to enhance grid stability. By facilitating a more diversified and dynamic energy market, they bring value to regulators, asset owners, traditional utilities, and consumers.

- With DC microgrids having a focus on local resource optimization at the low or medium voltage level, the introduction of VPPs creates new opportunities for market interactions within the ecosystem. Depending on its business model, a DC microgrid owner can for example engage directly with the newly formed markets, participate through a VPP, or position itself as a VPP.

5.3. *Intelligent microgrids*

DC microgrids are also set to be transformed by the integration of emerging technologies, such as the Internet of Things (IoT), artificial intelligence (AI), and blockchain.

- These advancements have the potential to greatly improve the efficiency, reliability, predictability, security, and transparency of microgrids. By incorporating advanced sensors, data analytics, and real-time monitoring systems, microgrids can become more intelligent. AIoT-driven intelligence can facilitate the prediction of generation and consumption patterns, optimization of energy flow, and automation of decision-making processes, resulting in enhanced microgrid efficiency and reduced costs. This can help facilitate businesses in capitalizing on the value of data through data-driven services and platforms, and enabling servitization in the energy sector. Better insights, real-time decision-making and increased transparency for all participating actors aid in democratizing the energy sector, a feat that opens the door to the emergence of new players, business models and opportunities. As these technologies continue to advance, the potential for innovation and transformation in the DC microgrid industry is vast.

- Moreover, blockchain, which involves decentralized transaction verification, can be utilized to incentivize and empower prosumers to seamlessly trade power and make payments. Integration of blockchain technology allows for secure energy trading, enabling the creation of decentralized marketplaces and fostering greater energy independence. A blockchain-based system could establish a distributed system of trust, enabling a home with solar power, for instance, to provide energy to a neighbor during a power outage even when the central grid is disconnected. This would promote resiliency by establishing self-sustaining cells of energy production and consumption.

5.4. Supergrids

A supergrid (or super grid) refers to an energy distribution network that connects various sources of renewable energy, such as wind farms and solar installations, with consumers across large geographical areas. Supergrids aim to optimise the use of renewable energy resources by linking areas with high generation potential to areas with high energy demand. These networks are designed to efficiently transmit electricity over long distances, significantly reducing energy losses and enabling the integration of renewable energy on a massive scale. By connecting remote renewable energy installations to supergrids with minimal losses, this technology enables the exploitation of vast resources that were previously inaccessible due to transmission constraints. For example, offshore wind farms in remote locations with strong wind resources can supply electricity to major cities located hundreds of kilometres away through DC.

How the development of supergrids will affect the business models of DC microgrids will depend strongly on ownership and regulatory boundary conditions of supergrids. Structures and solutions identified as optimal for a power system in one country or region, may not be optimal in another, due to differences in base scenario (grid structure, generation mix, etc.), available natural resources, existing legislation, social acceptance, and the like. Furthermore, investing in a supergrid carries financial risks, making it unlikely for a single company to undertake such an investment. Instead, a consortium of large companies or governments is more likely to invest. This investment decision will impact the development and operation of the systems, as owners expect a return on their investment. The emergence of supergrids will influence the economics of DC microgrids as supply and demand will increasingly be matched across the supergrid, making microgrids more interdependent.

5.5. Regulation

The increased complexity of DC microgrid business models, the need for increased partnerships, and the introduction of conflicting value logics all point to the critical role that regulation will play in determining the future of this emerging technology. A key observation from the case studies was the influence of the national energy markets, and the regulatory vehicles in particular, in the feasibility and viability of grid-connected DC microgrids. For instance, the Dutch postcoderoos regulation scheme encourages energy communities by offering a partial tax exemption for owners of DERs in their postcode area or neighbouring postcode area. Community members invest in cooperatively-owned generation assets and receive a tax deduction for their share of power generated, an incentive afforded predominantly to co-operatives and resident associations rather than corporations and private companies. Regulation can also be used to remove barriers to entry for new market participants, create a level playing field for all stakeholders, and ensure that the benefits of DC microgrids are shared equitably. The extent to which regulation will accommodate value logics that have traditionally been at odds with market regulation practice will influence citizen participation and prosumer dynamics. In all case studies, fitting within regulatory frameworks that were not designed for DC microgrids was paramount.

6. Conclusion

In conclusion, this chapter has provided an overview of DC microgrids and their significance in the energy transition. It has discussed the paradigm shift from centralized energy distribution to a decentralized and sustainable approach, highlighting key changes in the value chain including the emergence of prosumers and the de-commoditization of electricity production. The chapter has emphasized the importance of business model innovation to bring DC microgrids to market and has identified potential options based on ownership. Furthermore, it has presented examples of use cases and discussed future perspectives, such as dynamic boundary microgrids and expected technological trends. Overall, this chapter contributes to the understanding of DC microgrids and the potential business models that may shape the future of the energy industry.

The case studies shed light on the viability of diverse business models and implementation approaches for DC microgrids, emphasizing the importance of considering various ownership structures, contracting methods, and evaluation metrics when designing and operating such systems. The key takeaways from the case studies are

- Firstly, DC microgrids offer a multitude of benefits, including improved energy access, increased energy resilience, reduced energy costs, and enhanced grid stability. Depending on the chosen business model, DC microgrids can also provide valuable resources to the Distribution System Operator (DSO) and the wholesale system.
- Motivations for implementing DC microgrids varied, with some aiming to provide access to energy in developing regions where a significant portion of the population remains unconnected to the grid, while others sought reduced energy costs and energy independence in developed regions with reliable grid infrastructure.

- A common challenge across most case studies was the regulatory barrier that hindered the integration of customer Distributed Energy Resources (DERs). As a result, the business models devised often relied on regulatory workarounds to ensure the successful implementation of DC microgrids.

- Successful DC microgrids frequently involved collaboration among multiple stakeholders, including utilities, businesses, and community organizations. The case studies underscored the importance of considering local conditions, regulations, and stakeholder preferences when designing and implementing DC microgrids, as well as the need for effective customer engagement.

Overall, the case studies demonstrate the significant potential of DC microgrids to deliver a wide range of benefits, including improved energy access, cost savings, and enhanced resilience. However, they also highlight the necessity for careful planning, collaboration among stakeholders, and regulatory considerations to ensure successful implementation.

References

Afuah, A. and Tucci, C. L. 2023. Internet Business Models and Strategies: Text and Cases, volume 2. McGraw-Hill New York.

Aisa, A., Sharata, A., Almoner, M. et al. 2022. Modeling and simulation of a hybrid power system for hospital in sabratha, libya via homer software. In 2022 IEEE 2nd International Maghreb Meeting of the Conference on Sciences and Techniques of Automatic Control and Computer Engineering (MI-STA), pp. 161–166. IEEE.

Ariemu, O. 2023. Nigeria loses $26.2bn yearly to unreliable power supply- Nwangwu — dailypost.ng. https://dailypost.ng/2023/10/11/nigeria-loses-26-2bn-yearly-to-unreliable-power-supply-nwangwu/, 2023 [Accessed 26-12-2023].

Barnes, J., Hansen, P. and Darby, S. 2022. Smart and Fair Local Energy: Introducing Energy Local, Oct. 2022. URL https://doi.org/10.5281/zenodo.7147446.

Brown, D., Hall, S. and Davis, M. E. 2019. Prosumers in the post subsidy era: An exploration of new prosumer business models in the uk. Energy Policy, 135: 110984. ISSN 0301-4215. URL https://www.sciencedirect.com/science/article/pii/S0301421519305713.

Cai, X., Xie, M., Zhang, H., Xu, Z. and Cheng, F. 2019. Business models of distributed solar photovoltaic power of china: The business model canvas perspective. Sustainability, 11(16): 4322.

Chesbrough, H. 2007. Business model innovation: It's not just about technology anymore. Strategy & Leadership, 35(6): 12–17.

ElMaamoun, D. 2021. A Market Analysis of Virtual Power Plants and Some Ideas for Potential Opportunities. PhD thesis.

Garip, S., Bilgen, M., Altin, N., Ozdemir, S. and Sefa. 2022. Reliability analysis of centralized and decentralized controls of microgrid. In 2022 11th International Conference on Renewable Energy Research and Application (ICRERA), pp. 557–561.

Geissdoerfer, M., Pieroni, M. P., Pigosso, D. C. and Soufani, K. 2020. Circular business models: A review. Journal of Cleaner Production, 277: 123741.

Glasgo, B., Azevedo, I. L. and Hendrickson, C. 2016. How much electricity can we save by using direct current circuits in homes? understanding the potential for electricity savings and assessing feasibility of a transition towards dc powered buildings. Applied Energy, 180: 66–75.

Hansen, P. and Barnes, J. 2021.Distributed energy resources and energy communities: Exploring a systems engineering view of an emerging phenomenon.

Hirsch, A., Parag, Y. and Guerrero, J. 2018. Microgrids: A review of technologies, key drivers, and outstanding issues. Renewable and Sustainable Energy Reviews, 90: 402–411. ISSN 1364-0321. URL https://www.sciencedirect.com/science/article/pii/S136403211830128X.

IEA. 2015. India energy outlook - world energy outlook special report 2015.

Islam, M. M. M., Kowsar, A., Haque, A. M., Hossain, M. K., Ali, M. H., Rubel, M. and Rahman, M. F. 2023. Techno-economic analysis of hybrid renewable energy system for healthcare centre in northwest bangladesh. Process Integration and Optimization for Sustainability, 7(1-2): 315–328.

Jhunjhunwala, A. 2017. Innovative Direct-Current Microgrids to Solve India's Power Woes — spectrum.ieee.org. https://spectrum.ieee.org/innovative-direct-current-microgrids-to-solve-indias-power-woes, 2017 [Accessed 26-12-2023].

Kahn, M. and Longcore, T. 2014. A feasibility analysis of installing solar photovoltaic panels over california water canals. UCLA Institute of the Environment and Sustainability, Los Angeles, CA.

Karami, M. and Madlener, R. 2018. Business model innovation for the energy market: Joint value creation for electricity retailers and their residential customers.

Kaur, P., Jain, S. and Jhunjhunwala, A. 2015. Solar-dc deployment experience in off-grid and near off-grid homes: Economics, technology and policy analysis. In 2015 IEEE First International Conference on DC Microgrids (ICDCM), pp. 26–31. IEEE.

Kober, T., Schiffer, H. -W., Densing, M. and Panos, E. 2019. Global energy perspectives to 2060 – wec's world energy scenarios 2019. Energy Strategy Reviews, 31: 100523, 2020. ISSN 2211-467X. URL https://www.sciencedirect.com/science/article/pii/S2211467X20300766.

Koreneff, G., Similä, L., Forsström, J., Rinne, E., Schöpper, C., Segerstam, J., Repo, S. and Kilkki, O. 2022. Local flexibility markets: Smart otaniemi d4. 1. 2020.

Mauger, R. 2022. Defining microgrids: from technology to law. Journal of Energy & Natural Resources Law, pp. 1–18.

Mitchell, D. and Coles, C. 2003. The ultimate competitive advantage of continuing business model innovation. Journal of Business Strategy, 24(5): 15–21.

Newell, R., Raimi, D. and Aldana, G. 2019. Global energy outlook 2019: The next generation of energy. Resources for the Future, 1: 8–19.

Nguyen, N. -H., Le, B. -C. and Bui, T. -T. 2023. Benefit analysis of grid-connected floating photovoltaic system on the hydropower reservoir. Applied Sciences, 13(5): 2948.

Olivares, D. E., Mehrizi-Sani, A., Etemadi, A. H., Cañizares, C. A., Iravani, R., Kazerani, M., Hajimiragha, A. H., Gomis-Bellmunt, O., Saeedifard, M., Palma-Behnke, R. et al. 2014. Trends in microgrid control. IEEE Transactions on Smart Grid, 5(4): 1905–1919.

Osterwalder, A. 2004. The business model ontology a proposition in a design science approach. PhD thesis, Université de Lausanne, Faculté des hautes études commerciales.

Osterwalder, A. and Pigneur, Y. 2010. Business Model Generation: A Handbook for Visionaries, Game Changers, and Challengers, volume 1. John Wiley & Sons.

Peter Asmus, L. V. and Forni, A. 2018. Microgrid Analysis and Case Studies Report - California, North America, and Global Case Studies— energy.ca.gov. https://www.energy.ca.gov/publications/2018/microgrid-analysis-and-case-studies-report-california-north-america-and-global [Accessed 25-12-2023].

Premium Times. 2021. Fashola commissions electricity project that could save nigeria n270 million. https://www.premiumtimesng.com/news/more-news/471963-fashola-commissions-electricity-project-that-could-save-nigeria-n270-million.html?tztc=1 [Accessed 26-12-2023].

Roser, M. 2020. Why did renewables become so cheap so fast?, Dec 2020. URL https://ourworldindata.org/cheap-renewables-growth?qls=Q MM_12345678.0123456789.

Sadekin, S., Zaman, S., Mahfuz, M. and Sarkar, R. 2019. Nuclear power as foundation of a clean energy future: A review. Energy Procedia, 160: 513–518.

Sharma, S. K., Palwalia, D. K. and Shrivastava, V. 2019. Performance analysis of grid-connected 10.6 kw (commercial) solar pv power generation system. Applied Solar Energy, 55: 269–281.

Timmers, P. 1998. Business models for electronic markets. Electronic Markets, 8(2): 3–8.

Ton, D. T. and Smith, M. A. 2012. The us department of energy's microgrid initiative. The Electricity Journal, 25(8): 84–94.

UN in Nigeria. 2021. UN Deputy Chief calls for collective action to end energy poverty as Nigeria launches Solar Mini-Grid Plant — un.org. https://www.un.org/africarenewal/news/un-deputy-chief-calls-collective-action-end-energy-poverty-nigeria-launches-solar-mini-grid [Accessed 26-12-2023].

Vanadzina, E., Mendes, G., Honkapuro, S., Pinomaa, A. and Melkas, H. 2019. Business models for community microgrids. In 2019 16th International Conference on the European Energy Market (EEM), pp. 1–7. IEEE.

Wang, R., Hsu, S. -C., Zheng, S., Chen, J. -H. and Li, X. I. 2020. Renewable energy microgrids: Economic evaluation and decision making for government policies to contribute to affordable and clean energy. Applied Energy, 274: 115287.

World Bank. 2021. Nigeria to Improve Electricity Access and Services to Citizens —worldbank.org. https://www.worldbank.org/en/news/press-release/2021/02/05/nigeria-to-improve-electricity-access-and-services-to-citizens [Accessed 26-12-2023].

Zott, C., Amit, R. and Massa, L. 2011. The business model: Recent developments and future research. Journal of Management, 37(4): 1019–1042. URL https://doi.org/10.1177/0149206311406265.

Index

L

Levelized cost of electricity 191, 196–198, 200, 204

M

Machine learning 78, 80, 86, 125, 136, 139, 140, 142, 211, 220, 221, 227, 228
Microgrids 29–31, 33–45, 103, 104, 106–108, 110, 112, 176–181, 183–185, 189, 200–205, 209–212, 217–228

N

Net present cost 191, 196–198, 203, 204

O

Optimization 175, 177, 179, 180, 200, 201, 204, 211, 217, 220, 228

P

Performance 211, 217–221, 227
Photovoltaics 70, 77, 78, 85, 89, 90
Program learning outcomes 159

R

Rechargeable battery 144–146
References 53

Reliability 210–212, 215, 217–225, 227
Renewable energy sources 30, 32, 33, 43, 44
Resources 2, 3, 19, 25

S

Safety 211, 213, 214, 216, 217, 219–222, 224, 225, 227
SCADA 88, 94
Security 50, 52, 54, 58, 60, 61, 63–65, 211, 212, 215, 217–220, 222–224, 227, 228
Smart contract 52, 54, 56–62, 65
Smart grid 104, 105
State estimation 121–126, 138–141, 144–146
State of charge 25

T

Triple helix 160

V

Verification 211–216, 219, 222, 224–227
Virtual power plant 103–107, 109, 110, 113, 117